Population Geography

Population Geography

Rodolfo B. Valdenarro

Delve
PUBLISHING
www.delvepublishing.com

Population Geography

Rodolfo B. Valdenarro

Delve Publishing
224 Shoreacres Road
Burlington, ON L7L 2H2
Canada
www.delvepublishing.com

Email: orders@arclereducation.com

ISBN: 978-1-77407-620-0 (Hardcover)

ABOUT THE AUTHOR

Rodolfo B. Valdenarro Jr., is a Professor and licensed education practitioner at Centro Escolar University, Manila. He served as Director of Student Affairs and Services at Laguna State Polytechnic University from 2016-2019. He finished Master of Arts in Teaching Social Science at University of Rizal System and presently taking Doctor of Philosophy in Southeast Asian Studies at Centro Escolar University. He teaches Professional Education subjects, Social Sciences (History, Sociology, Humanities, Anthropology, Economics, Geography, Political Science) and Methods of Research. He served as Contributor on the 2011-2016 Philippine Youth Development Plan (PYDP) and President of State Universities and Colleges- Association of Student Affairs Administrators in Region IV (CALABARZON and MIMAROPA), Philippines from 2018-2019. Mr. Valdenarro has presented researches on Social, Behavioral, Educational and Gender Studies in national and international conferences.

TABLE OF CONTENTS

LIST OF FIGURES

LIST OF ABBREVIATIONS

ANC	Antenatal Care
ARCADE	African/Asian Regional Capacity Development
ASFRs	Age-Specific Fertility Rates
CSISS	Center for Spatially Integrated Social Science
DHS	Demographic and Health Surveys
IUSSP	International Union for the Scientific Study of Population
LMICs	Low- and Middle-Income Countries
MAUP	Modifiable Area Unit Problem
MDGs	Millennium Development Goals
MIPAA	Madrid International Plan of Action on Aging
NCDs	Non-Contractible Diseases
NSNP	National Safety Net Program
OADR	Old-Age Reliance/Dependence Ratio
PICMME	Provisional Intergovernmental Committee for the Migration of Migrants from Europe
POADR	Prospective Old-Age Dependency Ratio
RCM	Random Country Model
SMEs	Small and Medium Enterprises
SRSG	Secretary-General on International Migration
TFR	Total Fertility Rate
UN	United Nations
UNCTAD	United Nations Conference on Trade and Development
UNHCR	United Nations High Commissioner for Refugees
UNRWA	United Nations and Palestine Refugees Works Agency in the Near East

PREFACE

Population geography has obtained its roots of intellectuality since the mid-nineteenth century, it studies how the nature of places is directly related to composition, spatial variation in the distribution of populates and also population growth. Variation spatiality has been a concern to the geographer in the contribution of the distinctive study of population complains with the demographer, who studies population scientifically (childbirth patterns, merger, and annihilation) and not overly concerned in the pressure of spatial variation, leaving not also behind the migration in general. Study of geography within the field of population examination has continued to be of great significance to social science and other subjects, multi and interdisciplinary boundaries of disciplines increasingly becoming interested in population matters. The fields of studies include sociology, psychology, history, biology, economics, and also not leaving behind demography though they are intertwined. Population geography is concerned with the levels of fertility and birth rates in the region, advanced modeling of interregional growth of population and detailed patterns of mortality which includes disease diffusion. Nevertheless, movement and spatial abnormality than with other things are more interested for population geographers.

This particular book was prepared with utmost scrutiny of different researches, journals, articles, and other different population geography related books. The main goal for preparing this book is to offer knowledge-based information to all readers passionate about population geography. There are several fields where this book can be used, for instance, population geography, economic geography, demographic transition, social geography, health geography among others.

The book has been divided into 10 chapters and each chapter focuses on a particular topic under population geography: Chapter 1 discusses about introduction to key concepts of population geography; Chapter 2 is about patterns of population composition; Chapter 3 highlights information about population aging; Chapter 4 presents topic about human migration; Chapter 5 presents information about global distribution of the human population; Chapter

6 discusses about population diversity in the seven continents; Chapter 7 offers detailed information about urban, urbanization, and human population; Chapter 8 is about information on population and development; Chapter 9 highlights topics about women and social development; and lastly, Chapter 10 is all about men and social development. All the ten chapters were comprehensively prepared for the readers to have a glimpse of population geography. The author paid more attention on the simplicity of the content in order to make the book comprehensible and understandable to all readers.

1

INTRODUCTION TO KEY CONCEPTS OF POPULATION GEOGRAPHY

CONTENTS

1.1. INTRODUCTION

John Snow's disreputable water pump handle discovery marked its 150th anniversary in September 2004. A chronic outbreak of cholera was witnessed in England's capital, London on 31 August, 1854. Reckoning that the illness in question, cholera, was waterborne and very contagious, John Snow mapped the disposition of deaths due to cholera that surrounded the Broad Street water pump which he highly suspected to be the disease's source. In 1849, Snow conducted a review of all area residents' death records and went ahead to interview their relevant household members. These event series helped him determine that the majority of the folks who succumbed to cholera lived around and had drunk from the water pump at one point. John then presented his discovery to the community leaders resulting in the elimination, or rather removal of the said water pumps on 8 September, 1854. This, of course, prevented further deaths due to cholera. Often, John Snow is regarded as a legend in the field of epidemiology; having been one of the first people to employ epidemiologic techniques in the study and resolving of factors affecting a population's health and illness. Thanks to his study, substantial advancement was witnessed in terms of preventative actions and measures in the interest of collective health. Additionally, Snow is also considered a legend figure having provided and employed one of the first-ever examples of map power in the revelation of demographic undertakings and spatial associations.

Some several other early researchers and scientists employed similar techniques of mapping spatial dispositions in the examination of demographic episodes. In 1861, Mayhew mapped crime prevalence in London alongside poverty, disease, and urbanization. In 1902, Charles Booth emerged as one of the prime researchers to map spatial and social distance as per his survey of neighborhood dissimilarities in London. During the same period, Florence Kelley also conducted a homogeneous study by mapping penury in the Chicago slums. She learned, from her study, that several families frequently shared a single housing unit. Florence Kelley was one of the first individuals to employ cartograms in the representation of family density in terms of housing (CSISS, 2008). As indicated by these studies and research, maps feature quite enormous analytical capabilities. In past times, maps have played very vital roles in the proper and detailed understanding of various demographic occurrences among their characteristics. Geographic epistemology is typically represented by maps-spatial thinking. These are some of the scientists and researchers who were pioneers of spatial demography and population geography.

Population variations, characteristics, as well as spatial distribution are all addressed by population geography. The terms spatial demography and population geography are often used interchangeably. However, to be more precise, population geography can be explained and classified as a sub-disciplinary domain of geographic scrutiny that specifically deals with population processes and demographic issues in a spatially-explicit approach, anchored on the correspondence between places and people. Over the last few decades, substantial emphasis and attention have been drawn to the significance of a spatial perspective when tackling demographic research and any relative studies. Though not the only, geographers were among the first population scientists and researchers to embrace and endorse the significance of spatial thinking. Currently, population geography is considered a sub-discipline in the field of Geography as well as Demography.

1.2. HUMAN POPULATION

Is it possible to imagine a universe with the absence of human beings? In this scenario, who or what would have exploited the available resources to form, or rather create the cultural and social environment? Human beings play a very vital role in the development of the economy, and society at large. People are responsible for the creation and even the utilization of resources. As a matter of fact, people themselves are resources and only vary in quality. Take coal for example; it is but a type and piece of rock. That is until human beings invented the technology used to obtain it and turn it into an important 'resource.' Various natural occurrences such as flooding of rivers or even tsunamis only become 'disasters' once they affect an inhabited village or town.

This, therefore, has a population as the focal component of social studies. Population is also the reference point from which every other single element is observed and derives meaning and significance. Only when related to humans, disasters, resources and calamities all gain meanings. Human distribution, numbers, characteristics or qualities as well as growth all form the fundamental background for comprehending and valuing every other aspect of the environment.

Figure 1.1. Human Population.

Source: https://www.npr.org/sections/goatsandsoda/2015/08/07/429768917/11-billion-people-by-2100-and-india-will-be-more-populous-than-china.

As a fact, human beings are the prime and sole producers as well as consumers of various earths' resources. It is therefore pretty important to determine how many people there are in a staid country, where they live, the rate and reason for their population increase, or decrease as well as their traits and characteristics. The census of a specific country provides all the information needed regarding the entire population of a country, living situations and the location where each person usually lives. There are three principal questions relative to a country's population, these includes:

- **The Size and Distribution of a Country's Population:** How many individuals reside in the country and where specifically are they located?;
- **The Growth in Population and Undertaking of its Change:** How has the country's population changed, or rather grown over a period of time?; and
- **Qualities or Characteristics of a Said Population:** What are their features including sex-composition, age, literacy degree, state of health as well as structure of occupation?

1.3. POPULATION GEOGRAPHY DEVELOPMENT

The origin of population geography is traced by Lesek Kosinski from the schools of human geography in French and German of the early 20th century and half of the 19th century. The schools had a specific concern on the relationship between population and the environment and also population mapping. Nonetheless, it was only after World War II that Pierre George (1951) came up with a publication (mirroring that country's specific fascination in demographic matters), and Glenn Trewartha (1953) where the sub-discipline started to take its modern shape in the United States. In Germany, population geography had been disparaged because of its cooperation with Nazi policy. In addition, few other countries were slow to pursue even though India, Japan, and the Soviet Union made significant progress. For Glenn Trewartha, "all other elements are viewed and from which they all singly and unitedly extract connotation and meaning from the population as a point of quotation. It is a population that fixes up the focus." His "organization and content structure was tentative." For population geography field was broadly defined, including the dynamics of population growth, historical population geography, population distribution, migration, socioeconomic characteristics, and population structure. Geography was a very strong interest with the international pattern of population dissemination at the core of the population and the way demographic processes were reflected into a broad physical and human environment.

The heightened opening of demographic information and the very importance of population concerns in both the advanced and the developing nations facilitated the postwar growth in this field. Population geography was given a solid space in the geography syllabus in many countries after the publication of many persuasive textbooks—such as those by George Demko, Harold Rose, and George Schnell (1970) and John Clarke (1965). The strengthening of the field came as a result of an improved institutional environment marked by the activities sponsored by the International Geographical Union commission (exclusively from the late 1950s, the Association of American Geographers Specialty Group (from 1980) and the Population Geography Study Group of the Institute of British Geographers (from 1963). The involvement of population geographers with multidisciplinary bodies such as the Scientific Study of Population of the International Union has been more limited.

Population geography early foundations were quite divergent from demography (and certainly had comparatively barely effect on), but from the 1970's it was with acceleration disputed that demographic methods were essentially to be focused on more apparently. Hence, publications such as Robert Woods (1979) shed outstanding prominence to the central demographic development of mortality, virility, and rather less to the movement. The concept was to combine spatial demography and population geography around the center of approach derived from demography. This corresponded with the immense use of measurable methods in geography generally. The publications such as Allan Wilson and Phillip Rees (1977) which focused on proportional accounts usage and models analysis of spatial demography and Peter Batey's and Peter Congdon (1979) brought an interdisciplinary look on "zonal demography."

The formal onset of population geography as sub-specialty of Geography is owed to Glen Trewartha's 1953 presidential address at the Association of American Geographers' annual meeting, at which time; geography was generally divided into cultural geography and physical geography. On the said occasion, Glen introduced a geographic triage comprising the cultural earth, natural earth, as well as population. In relative modern terms, this triage includes the cultural environment, the physical surroundings, and people. In his alignment, he had 'population' on top then cultural and physical environments at the bottom as illustrated by Figure 1.2.

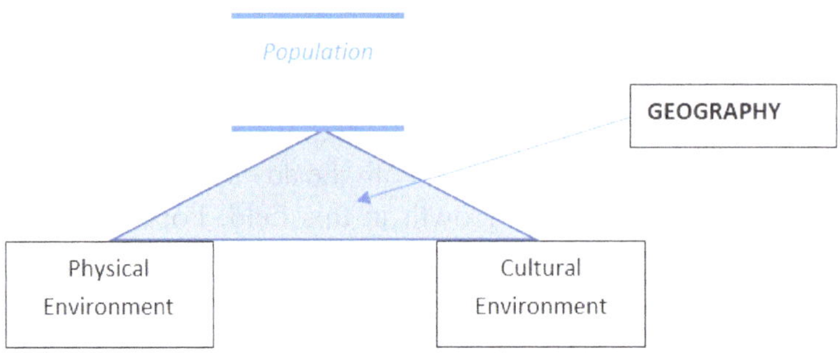

Figure 1.2. Geographic triage.

Source: Authors creation.

Trewartha was a climatologist as well as a physical geographer. On the mentioned occasion, Trewartha was calling for geographic sciences to be viewed from an anthropomorphic point of view. Both physical and cultural studies were made significant, so as the relevance of population. He was endorsing and encouraging the study of populations scientifically, especially due to their correspondence with the physical environment. Glen mentioned and emphasized that every other single element derived their meaning only when observed from a 'population' point of view. He added that population is the one responsible for furnishing the focus.

Geographical texts and scripts have been known to recognize Glen Trewartha as the one responsible for the initiation of population geography even though it hasn't really become the zenith of Geography as a discipline, as proposed by Trewartha. As much as Trewartha has gained much fame on this subject, understand that he wasn't really working alone in establishing population as an independent scientific study field. One of the most influential classic publications on demography, "The Study of Population: An Inventory and Appraisal," authored by Duncan and Hauser in 1959 comprises of a whole chapter on "Geography and Demography." This clearly establishes Population Geography as a prototypical interdisciplinary sphere of the duo.

Among the few scholars who took it to themselves to explain in-detail the comprehensive context of Glen's address is Weeks (2004). Trewartha attended the University of Wisconsin. This institution was, back then, and is still one of the most vibrant avenues for the population studies. This is because the call for 'population' to be a scientific study constantly echoed throughout the institution. What more is that this call gained significant attention and was being heard. It is during this period that the Rockefellers the Population Council, which is the supreme sound currently in policy-oriented studies. Moreover, the International Union for the Scientific Study of Population (IUSSP) alongside the United Nations (UN) was laying the groundwork for a very crucial event on the same, the World Population Conference, thereafter hosted in Rome in 1954.

Throughout the 1960s, all the way to the 1970s, population geography thrived and, in some way, flourished, and matured in the 80s. By this time, population geography had become subjugated by quantitative techniques and spatial analysis. It had become a substantial Geography sub-discipline whose province was population studied scientifically. Momentous, in this evolution, was the role of Wilbur Zelinsky in his bid to distinguish a demographer from a population geographer.

Through the 1970s, the scientific approach of studying geography influenced and affected other spheres of Geography. During the late 70s, a sub-discipline of medical geography was established. In 1977, an article was scripted on the same by Melinda Mead entitled "Medical Geography as Human Ecology: The Dimension of Population Movement." This article is often quoted as the linchpin of medical geography's establishment as a sub-discipline in geography. What's quite interesting about this article is that Melinda included a diagram quite similar to Trewartha's triage; nonetheless, it spotlights health as the sequel of the potent correspondence between the three facets of culture, environment, and population. Additionally, this empirical study of the population went ahead to become a major concern of other sub-spheres including regional science and economic geography. Hagerstrand argued that space and time were not unconventional of each other. Hagerstrand's contribution to the course was as influential as it was eloquent. His studies and research were fixated on innovation diffusion as he researched on how new technologies spread. Individuals, together with their ideas budge as one. Antecedent to his opuses, population geography, and regional science did recognize the significance of both time and space. However, they tended to keep both discrete analytically. The work done by Hagerstrand was formidably influential in both Regional Science and Geography. This is especially because his work sited human beings in space and time. Position people in place and time are fundamental to extant population geography.

1.4. THE NOTION OF POPULATION GEOGRAPHY

As per Philip and Ogden, population geography is focused on the comprehension of the regional distinctions in the universe's covering of human beings. Furthermore, Population geography had been earlier defined as a discipline of geography that research on human population distribution over the mundane surface.

Geography is responsible for the facts and phenomena distribution over the earth's surface. In a more palpable sense, geography can be described as the study of the earth's surface and of any occurrences that affect it mutually.

Similar to many other fields, Population Geography has also had a pre-scientific phase as long as the history of human beings. People have always been curious of their neighbors, their neighbor's organization, or structure, as well as what to fear or expect from the lot.

Discovery of America, as well as the eighteenth and nineteenth-century explorations, enabled us to become aware of the actuality of some ignored cultures and towns. As time went by, the study and observation of diverse populations occupying the earth, as well as their organization in terms of living and exploiting their respective territories paved the way for the unearthing of multiple scientific disciplines including human geography. It was also argued that population geographers must be involved in the concoction of census among any other conventional population data assemblage in a bid to ensure that appropriate and versatile spatial systems are adopted and employed. It was put across that in most countries, specifically in Asian countries, the adopted systems didn't enable data availability for spatial denominations which are economically, environmentally, and socially meaningful.

Ecology is the study of a live entity in its natural environment. In their state of living, human beings are unwrapped in a definite natural ambiance, for which their natural ecology exists. However, unlike in the case of other species, bonds presented between human beings and their means aren't definite.

Below are the three main facets of human population in which population geography is mainly concerned:

- **Distribution and Size:** This also includes the rural-urban dissemination of population;
- **Population Structure and Composition:** This comprises of a set of economic features, i.e., structure of the workforce, rate of workforce participation, etc.; Social characteristics, i.e., ethnicity, linguistic population composition, literacy levels, etc.; demographic attributes, i.e., marital status, sex structure, age, etc.; and
- **Population Dynamics:** This includes the present and past trends in development as well as its spatial reflection in terms of various population change components, migration, mortality, and fertility.

1.5. WHY STUDY GEOGRAPHY?

According to some researchers, here are the top 10 reasons behind the study of geography:

- To gain proper understanding and comprehension of some of the principal physical systems that influences our everyday life. This includes wind currents, water cycles, relationships between the earth and sun, etc.;
- To acquire knowledge on the location of various places most especially their cultural and physical attributes to helps one function more constructively in this world which grows more and more interdependent every other day;

- To grasp an understanding of geography's history as well as its significant role in the evolution of man, his environment, places, and also ideas;
- To help them develop a mental map of one's community, territory, nation, and the universe at large. This should help them easily comprehend the "where" of any events or places;
- To be able to explain how various physical systems and human activities have aligned and even changed the earth's surface;
- To fully understand and comprehend the spatial arrangement of society and also realize the order in whatever usually seems to be the random dispersion of human beings and places;
- To identify spatial distributions, both on a local and universal scale, for the proper comprehension of humans and places complex connectivity;
- To enable proper and sensible judgment calls in matters and situations involving society-environment relationships;
- To grant people the proper appreciation of planet earth as their homeland and also gives insights on the wise management and use of its resources; and
- Of course, studying geography could help one properly understand universal interdependence and to become a wiser and generally better universal citizen. Migration happens to be one of the biggest challenges being faced by population geography.

1.6. WHAT EXACTLY IS HUMAN MIGRATION?

In the simplest terms possible, migration can be described as the movement or maneuver of people or communities from one physical location to another, entailing either temporary or permanent settlement schemes (Figure 1.3).

Figure 1.3. Migrating Human.

Source: https://www.nationalgeographic.org/activity/introduction-human-migration/.

One societal aspect that is quite noticeable today is the rate of rural-urban migration. While the population is steadily increasing in urban centers, it is in a steady dive in rural areas. These are some of the reasons for the common concern on how to slow, or rather reduce rural-urban migration. However, policymakers are more concerned that the high urbanization rates, as well as urban unemployment, may pose quite strong political and social challenges on the reduction of rural-urban migrations. This segment looks to highlight some of the challenges facing rural-urban migrations.

Also provided in this content are some of the migration theories providing theoretical reasoning backing rural-urban migration. This opus captures from factual literature studies review of rural-urban migration from various secondary information sources as well as the internet.

1.6.1. Reasons Behind the Movement of People

According to various authors of different sources, among the reasons why people move, include physical, cultural, economic, and political reasons, as well as subsets or combinations of these reasons (for instance, migration due to political reasons may include cultural reasons as well, such as religious persecution).

People move in a bid to improve or develop; basically, the developments or improvements being referred to include living standards, safety, resources, and economy, religious, and also political reasons. For most Americans, the reasons they move include retirement, family connections and also job relocations and transfers. However, for the case of immigrants coming into various benefiting nations, America included, reasons behind their movement are predominantly safety, political justice, or economics. Either way, it is common knowledge that the leading reason behind people's relocation in the current world is due to safety purposes. Regardless of whether the reasons for moving are genocides or hurricanes, safe living environments are man's number one concern. Keep in mind that natural disasters such as landslide, earthquakes, hurricanes, and tornadoes among others are quite big reasons for people's relocation. Apart from nature, a person's safety can also be threatened by another individual. Genocides, war among other phenomena that threaten life can also lead to migration. In such cases, people flee from their own homes and territories to other much safer locations in fear of their own lives and those of their beloved.

Another reason for people's relocation is resources and economics. When an individual is struggling so much to fend for themselves and their family members, they search for new and better opportunities. In most scenarios, this may result in temporary or even permanent relocation. For one to come across better opportunities to fend for themselves, they have to relocate to a place where the economy is much stronger. This, in turn, results in sizeable incomes enabling them to cater to all their needs and those of their loved ones.

In most cases, the immigrant relocates to the country in question just long enough to come up with something substantial. However, it is quite common for them to move their household members to the said country and seek permanent citizenship. As stated above, another reason why people move is resources. This is quite self-explanatory. If people happen to lack water, food or any other essentials, they relocate to different places where these basic needs and resources are easily accessible. People whose homeland is stricken by drought or famine move to places further away to where water and food can be easily accessed.

The final key reason behind the relocation of people is religious and political complications. Of course, if one lives in a place where governmental policies and ways negatively affect their way of life, they'd want to migrate or move. Nations featuring unjust, or rather inhumane political policies and ways have numerous emigrants fleeing to other countries with more suitable environs.

Another reason behind migration is religion. Religious acceptance and religious persecution are key points in migration matters. People will always flee places or countries where they face torment and persecution because of their beliefs and settle in areas where their beliefs are acceptable.

As observed, people migrate due to various reasons. Nonetheless, most, if not all of the reasons stated above fall into one of these three key categories: politics and religion, safety, and living and finally economics and resources.

1.7. SOCIAL CLASSIFICATIONS IN POPULATION GEOGRAPHY

More than ten years ago, Jackson, and White (1995) urged population geographers to become involved in developing debates within the non-specific discipline. Included in their more powerful arguments was the necessity to consider the erected structure of social categories as well as the material interests inevitably served by those social categories.

Relatively, researchers have been quite uncritical of the techniques in which raw data is presented in ready-classified forms, and haven't exhaustively examined non-essentialist or alternative ways of approaching social classifications and labels.

Their suggestion was that demographic research and studies could be enhanced by staying clear of fixed categories that are so easily accessible in various data sources including surveys and census, and so commonly adopted and employed by population researchers. They also argued that for the cornerstone of population geography to be broadened, scholars need to take subjective individuals' identity meaning into account, and also how they are reconfigured and crafted in various social, economic, and political contexts.

Ever since the publishing of the above opus, there have been significant and notable, developments witnessed. McHugh (2000) unveiled the various ways and techniques in which ethnography functions towards proper understanding and comprehending affairs of belonging and identity creation at various scales of analytics. This is but one example of the same. Others include Graham's (1999), Silvey (2004) as well as Boyle (2002). All these scholars lay emphasis on the widening of the sub-discipline and serve to guide us away from various analyses of population organizations that are gender-blinded. Additionally, their works help us catch up with opuses of other geographers who are also actively involved in the feminist theory proceedings.

Still, there is a load of work yet to be executed in population geography studies in order to out-scale the essentialist analysis frame. Race especially keeps on being an ignored social classification within the literature and frames of population geography.

Even as the two scholars, Jackson, and White called for researcher's engagement in ongoing ideas on social construction, or rather the creation of race, and also the racialization of politics and policies, unsubstantial progress on the same has been witnessed in the last decade. However, note that population geographers aren't the only ones in this case: as recently argued by Nash (2003), Anderson (2002), Peake, and Schein (2000), as well as Kobayashi and Peake (2000), there hasn't been much done by most geographers in attending to anti-racist geography. There is a vehement cry for more and more geographers to get involved in critical race theory.

Racialized analyzation of population processes is way complicated as part of the mere employment of contemporary ethnic/racial categorizations and groupings. Note that societies are affected deeply by race in ways that aren't usually easily categorized, separated, or even identified. In understanding and comprehending the impact of race, try to identify and tease apart the multiple impacts of categorization while still staying conscious of the constant transformation of this categorization. Omi and Winant (1994) argued that in the discourse of racialization, attention is paid to the ideologies that cardinalize the racial categorization exercise as well as the material manifestations and consequences of docketing population classifications in specific ways.

For example, the typical demographic approach to ethnicity/race might be concentrating on residential segregation indices and spatial patterns. Still, an alternative racialized approach might be relating specific spatial and social practices to the broader social structure embedding those practices and from which their values and meanings are procured; that is, the institutions, hierarchies, and shifting race and racialization ideologies and their relation and reproduction to the existing economic, political, and social structures. Now, instead of reckoning on the 'mapping' of ethnic/racial divisions through the employment of census databases on a broader scale, some scholars including Craddock (2000) and Anderson (1989, '88, '89) disclosed how, from history, the Chinese turned racially classified during the spatial segregation of the Chinatowns where they were based.

Deconstruction of social categories and classifications is more likely to be especially strenuous in population geography. This is because this is a discipline where aggregate and data-driven thinking techniques still influence radical conceptualizations of race, generation, as well as gender. Understand that as much as conventional techniques and data sources are quite useful in the analysis and description of the character and form of present-day immigrant settlement in various locations, they still accomplish little in matters pertaining to the in-depth comprehension of:

- How class, location, and ethnicity/race shape the ethnic/racial identities of immigrants bound to sub-urban areas; and
- The contribution of suburban immigrant settlement to the ever-shifting conceptions of the major ethnic/racial boundaries.

In order to recognize the process that interlinks suburbanization, immigration, and racialization, you have to do away with the old techniques and adopt new ones such as participant observation, in-depth interviews, focused group discussions and landscape analysis. This approach of mixed techniques answers Wright et al., call for more crucial studies supplants 'whiteness' as the sole endpoint in assimilation orientations. It also serves as the beginning of articulation of how the suburbs become much more than a mere backdrop for ethnic/racial identity negotiations.

All the same, as much as these qualitative methods tend to be useful in providing answers to specific questions, it happens that focus groups provide way different perspective in trying to grasp ethnic/racial categories, stereotypes, attitudes, and their spatial and social construction. Via focus groups, experiences of both immigrant newcomers, as well as those of the non-Hispanic, native-born white residents were uncovered. Through the same focus group technique, the roles each play in the construction and classification of social groups as well as minority groups was also explored.

In the focus group deliberations, attendees pronounced both disagreement and agreement, highlighting consensus and unanimity in agreement cases and hence pinpointing group norms. Since ethnic/racial identity and ethnic/racial attitudes are both spatial and social creations strongly influenced by an individual's group affiliation and interaction, the setting of the focus group availed alternative strategies to hit the nail on the head. In cases of disagreements, subjects had to give an explanation, justification, and perhaps position modifications, availing insights behind their way of thinking. Also, a couple of subjects whose identities were censored were used to unveil some social procedure underlying knowledge articulation. This comprises stigmas associated with specific information or opinions, particularly regarding spatial and social boundaries erected and endorsed in the racialization processes. Consequently, the group sessions helped in the generation of new relative questions as well as verification of antecedent results obtained through different research channels. Therefore, in answering Jackson and White's (1995) call for population geography that is (re)theorized, one has to go beyond the scopes of the erected techniques and norms of the sub-sphere, as well as incorporate a mixed technique approach with the inclusion of focus groups.

1.7.1. The Use of Focus Groups

For any population geographers looking to uncover the formation of social categories, adoption of social categories had proved useful due to several reasons. The group provides a subject with a private space/arena in which he/she can define their identity. They are also able to occupy or associate themselves with different categories and establish new ones as well. The mutually exclusive classifications given in official surveys and documents adopt new significance and meanings once discussed in a congregation setting. Fixed identity conceptions such as impartial categorizations start to emerge. Additionally, identities situational natures, which are subject to change and highly context-dependent, begin to unveil. A focus group also plays a very significant role in the articulation of categories being a type of labeling, proving how powerful labels can be when hierarchically arranged and when used to 'include' as well as 'exclude.' Hence, by opening inquiries on population geography to comprise various identity notion and not the data categories presented by official statistics exclusively, a whole new level of complexity is introduced. Furthermore, also introduced are much more critical approaches to the influence of population processes by social classes.

For further articulation on the value of focus groups to population geographers, here are two other fundamental advantages of this technique:

- Great potential for spontaneous and unique group interactions; and
- Great potential for participants' empowerment.

1.7.2. Great Potential for Spontaneous and Unique Group Interactions

One of a very powerful benefit of adopting focus groups is that the interactivity between participants of the same provides an opportunity to unmask beliefs, opinions, and ideas that may not be unveiled even in the most in-depth survey questionnaire or even interview. Compared to the mentioned channels of data collection, a group setting does readily encourage spontaneous feedback and responses from the group participants. Keep in mind that in interviews, or say questionnaires, the moderator does impose a kind of discussion structure for the participants not to deviate from the topic at hand. However, in the case of focus groups, the discussions may easily deviate towards other issues that hadn't been earlier on considered by the researcher, but are germane to the topic at hand, and may even result in very significant findings.

As inter-participant interactions replace participant-researcher interactions, the emphasis is laid on the participants' perspectives. Focus groups allow participants to act as moderators, questioning, and making relevant explanations to one another. This, in turn, provides the researcher with a surrogate perspective on differences in experiences and also attitudes. This is owed to the fact that the participants themselves are responsible for various comparisons as well as explanations of their unshared perspectives. Additionally, focus groups not only hide or censor identities if need be, but they also advocate sharing of 'taboo' or rather private topics, as the more assertive group participants are bound to break the ice. Undeniably, the setting of a focus group may also subvert any particular meanings and prejudices brought to the project by the researcher.

An example of the significance of focus group interactivity in the production of both consensus and distinctive points of view is Secor's 2003 and 2004 research. Being interested in the various ways in which citizenship is met and also contended through everyday life's spatial practices, focus groups unveiled to Secor how some participants united in expressing how proud they were of their cultural identity; neighborhood and places of work. Yet, after being prompted by other members, the group began questioning their identity's unity and meaning. The group setting facilitated and opened up various discussions revolving around the subject of citizenship and belonging.

Once compared with isolated individual interview sessions, the advantages of group interactivity can be clearly articulated. In individual interviews, detailed information can also be acquired but through an entirely discrete context. In 1994, Wight employed both techniques in the exploration of sex thoughts, from an adolescent boy point of view. He observed that in individual interviews, the subjects modified their responses to suit the interviewer's morals and expectations. On the other hand, when it came to group settings, the boys tended to parody and exaggerate their different ideas and experiences to what they thought was required by the moderator and perhaps show off to the other participating boys. Consensus, thus, materializes when both individual interviews and focus groups are employed, however, in distinctive ways, illustrates the responsibility of geographical and social contexts in comprehending various phenomena. Group settings facilitate platforms on which participants can provide their sole interpretation of their behavior and attitudes in the group context. Thanks to a group setting's interactive nature, participating members are pushed to explain their theories and views on their attitude differences.

Certainly, a probable flaw of focus groups has to do with conformity. Focus groups findings can be undermined if the participating members happen to fall back to traditionalism. Bulow and Hyden (2003) highlighted that researchers need to pay attention to the participants' expression capacity. Dialogue analysis can inform the moderator of whether or not the participants have agreed, or rather developed a common ground on which they base their responses. An example of such a situation is when the participants begin responding to and building upon each other's responses. Inter-participant conversations can also reveal if they are all acting and responding as individuals. This is especially true if a participant focuses on their own experience without having to make connections to someone else's experiences. Depending on this kind of group identity build out, the attitudes may be irreducible to an individual, and will instead pose a group orientation. Consensus on interactivity when it comes to group data assemblage is tremendously valuable as a tool of research only if analyzed in a way that is attentive to a certain influence of the focus group setting.

1.7.3. Great Potential for Participants' Empowerment

Possibly, focus grouping can also incorporate marginalized lots, because the discussions of the group can empower members who usually have no 'voice' in discussing matters of their livelihood. These focus groups can also ensure that the precedence is given to the equivocator's knowledge hierarchy, their concepts and language as well as their universe's comprehension frameworks. Furthermore, participants can be empowered by earmarking focus group topics that are relevant or important to them. This is especially true whenever the participating members are presumed to be experts on the topic at hand.

More importantly, the inclusion of marginalized groups, or rather communities in studies and research on significant social matters can assist with the negation of feelings of systematic exclusion. There have been numerous researches conducted by various scholars that seek to empower the female gender, who otherwise felt like they were living on the society's margins. Pratt (1999, 2002) interacted with Filipina domestic workers and unveiled the group's subjection to exploitation, victimization as well as abuse. In the group's safety net, the participating domestic workers were able to stir the discussions to various topics that were important and significant to them. They were also able to interpret the specified researcher's topic deemed to be relevant to them. Additionally, focus group sessions with local service providers established and illustrated how certain discussions unfold as base practices in specific places, resulting in the establishment of women's spatial and social marginalization. Pratt worked cooperatively with the Philippine Women Center on various reports and papers in a bid to inspire more participants who experience the same situation, new dialogues and discussions.

This kind of joint efforts is not quite unconventional for scholars who employ focus group techniques. Certainly, these focus groups especially prove useful and resourceful for researchers who are committed in addressing subject matters that have the potential of engendering and stirring positive change. As defined by other research projects that employed focus group methodology, focus groups can help transform the comprehension of both participants and researchers in ways that new and also compelling.

Another benefit of employing focus groups is that it can help ease language barriers that primarily existed between the participants and the researcher. In instances of structured questionnaires and close-ended surveys, it is always assumed that the respondent is already aware of the language used to express the same ideologies. These barriers can be overcome with the researcher being able to listen to the conversation between participants in their native vocabulary or language. This idea speaks and acts on Jackson and White's suggestion on avoiding fixed categorizations. It can also create a researcher-research subject relationship that is less hierarchical.

You need to understand that none of these strategies are problem-free; however, they are worth considering negating participant's sentiments of exclusion from the system.

1.7.4. Focus Groups Design and Establishment

Kitzinger (1994), among other focus group advocates, explain that many scholars who've adopted and employ focus group methods don't justify their employment of this method adequately, nor do they detail the research design's specifics. As outlined by Morgan and Krueger (1993), the time and cost savings can lure scholars to adopt the use of focus groups without necessarily any factual ground for the technique in the topic, data objectives or even theoretical framework. Certainly, the logistic benefits of employing focus group sessions can present a fairly cheap and relatively simple means of sourcing quality data in a small window of time. However, get that time-cost efficiency often results in inadequate studies which may lead to shoddy reporting and poor design.

Perhaps the reluctance of population geographer in adopting and employing focus groups in conducting their research can be associated with the lack of high-quality viable examples. However, there are a few principles and considerations of the focus group technique that should see you surpass your expectations. Particularly, since the outcome of research entirely depends on the design of the research, here are three vital factors that should be considered when structuring a focus group. They include:

- Segmentation;
- The moderator's role; and
- Standardization.

Since focus groups can be employed for various research purposes, some other factors and considerations are bound to turn up depending on a specific project. Either way, mentioned are the three most fundamental factors that should produce massive success if properly addressed in advance.

1. **Segmentation:** In the simplest terms possible, segmentation can be described as a marketing term used to illustrate the purposive selection of groups and individuals according to specified characteristics or attributes. Due to the nature of research conducted through focus groups, it is quite paramount that the group composition is controlled and swayed in terms of the attendees' demographics and social traits. In general, these focus groups have to be analogous to the topic at hand as well as along certain attributes including age, social class, gender, and language among other characteristics.

Of course, increased segmentation tends to yield multiple groups; now, for the attainment of data saturation, it is recommended that the researcher completes around four to six groupings per a type with each group including around 6 to 12 active participants. The number of participants entirely depends on the research topic. In addressing a topic quite unfamiliar to most, researchers usually, select relatively large groups for them to obtain sufficient discourse. On the other hand, when tackling topics that are emotionally charged and quite familiar, the researcher has to restrict the number of participants to heighten the feeling of security and intimacy among the participants.

There are numerous advantages to segmentation when planning for a focus group on a specified research title. However, there are three key ones including:

- First and foremost, with respect to a specific attribute of the discourse topic of the focus group, homogeneity tends to intensify participants' sense of security result in active and vibrant participation.
- Secondly, segmentation ascertains that the ineffectiveness or dormancy of a specific group doesn't influence the outcome of the research.
- The third and final benefit of segmentation comes in once the focus groups have been conducted and concluded, and data analysis has commenced. Segmenting each group according to various attributes or features eases the researcher's coding ability of the transcripts across various categories of people and systematically.

These are just but the three main advantages to segmentation when looking to employ focal groups in various research and studies.

2. **The Moderator's Role:** Decisions on the moderator's role entirely depend on format preferred by the researcher(s); more open and exploratory formats or alternative standardized formats. Essentially, the moderator's key role in any focus group is to present the research topic as well as the end goals as the discussion commences. They are also responsible for ensuring that the group remains focused and the session progresses smoothly to the end. Often, this entails the introduction of new questions or/ and interrupting where they dim necessary. Additionally,

the moderator manages the dynamics of the focus group by curtailing some dominant discussion members and also encouraging the rather hesitant ones to become active.

As much as the importance and significance of the group's moderator are generally agreed, there are some arguments as to whether or not they require certain training and skills. When talking about skills and training, it doesn't necessarily mean that they should be professionals on the same. Some scholars argue that the best cause is to appoint a research member as the moderator since they are well-conversant with the research protocol, schedule, as well as the required coding scheme on completion of the focus group sessions. The verdict, in this case, is that a moderator should be adequately trained to easily create and maintain an environment where the collection of data is maximized following the project goals.

For perfect comfort, it is argued that the researcher matches various influential attributes of the participants to the moderator including age, social class, ethnicity/race, and gender depending on the topic of the study. There is a consensus that the focus group members feel less inhibited if they are of the same national, racial, and/or ethnic background with the session's moderator. This especially comes in handy when unveiling negative remarks or traits regarding other groups.

3. **Standardization:** It is described as the process through which researchers come up with similar procedures and questions across various groups. For the achievement of comparability across various groups, fixed, and pre-determined sets of questions have to be administered to every single focus group. Standardization will also enhance comparability when it comes to the analysis of the same. This is because of a consistent format and approach as employed and followed during each focus class's interactivity.

On the contrary, if the study features an explanatory structure, adjustment of format and research questions depending on each group's experiences may be preferred. This will, in turn, contribute to a collective set of information. However, note that this kind of open-natured and exploratory structure may be more consistent for those scholars not looking to impose the researcher's interpretations and assumptions on the study.

1.8. CURRENT GEOGRAPHY OF POPULATION

Although ideal publications such as that of Huw Jones (1981) restfully took a predominantly established perception for some geographers this connection to the demographical methods that beckoned an objectionable consolidation of population geography, separating it from the rest of the geography as the time when controversies around the demanding social approach, more calumniator look of entrenched information sources and approaches and steps to approximate as well measurable methods. Interestingly, still, their momentum came widely from the subject of geography their interests depicted ones declared within demography. These analytical geographers would concede with the anthropologist-demographer Susan Greenhalgh's examination that "what is lacking in the field is reflexibility on the politics of demographic praxis, the global political economies of the 1970s, the postmodernisms and post-coloniality of the 1980s and 1990s and even the perceptible force of the feminisms of any decade."

A closer look at potency and fatality rate has been aimed at specific to present the patterns of spatial dimension and their relationship with natural and communal conditions for example incidences of disease spatiality's, fertility, and or even mortality. Demography and geography have also been combined by others in coming up with the credible accounts of countries or continents. Demographers such as Ansley J. Coale and Susan Cotts Watkins have taken a concern in patterns of demographic change that have clarity on geographical dimensions both nationally and internationally, population patterns of fertility and mortality reconstruction on historical geographies as well as family unit and people formation through methods such as household reconstruction and the comprehensive guidance of the antiquity enumeration, enrollment, and ecclesiastical reports have been given to geographers by the historians.

Up until now, geographers involved in population matters have inclined most consideration to movement, aggregate estimation, and net progress at various scales: interregional flow building models: and figuring out commercial and communal causes and consequences. Migration considerations have comprised movements internationally: rural-urban, urban-urban, seasonal, perennial movements and intra-urban flows. Some geographers, for example, Paul White, Russell King, and John Connell have looked also at the idiosyncratic background of movement etching on exhaustive surveys and inventive literature.

Through the information and theories in population geography, they serve to make recall both to demographers and geographers with population passion in other subjects that demographic adjustments have spatial and also materialistic extents. At the same time, it makes geographers recall that population peculiarity is a very important element in the character of any region.

1.9. POPULATION CHANGE/GROWTH

There are three major population change processes including:

- Migration;
- Birth rates; and
- Death rates.

The variation between birth and death rates is referred to as the natural growth of the population. The birth rate is described as the number of births per a thousand people per year. It serves as a very vital growth component, since, in India, the rates of birth have been steadily higher than the rates of death.

On the other hand, death rates can be described as the number of deaths per thousand individuals per year. The Indian population has been on a constant growth rate due to the expeditious decrease in death rates over the years.

Until 1980, significantly high rates of birth and decreased rates of death led to substantially large variations between death and birth rates resulting in quite high population growth rates. As from 1981, the rates of birth also started dropping gradually, hence leading to a gradual decrease in the population growth rate. What are some of the reasons behind this trend?

The last population growth component is migration. Migration can be simply described as the movement or rather relocation of people across territories and regions. Migration can be classified as internal- within the same country- or international -from one country to another.

Unlike international migration, internal migration doesn't affect a country's population size but rather influences population distribution within the same country. Note that migration plays quite a crucial role in influencing the population composition and distribution.

In India, the population in towns and cities has steadily increased over the years thanks to rural-urban migration. In 1951, the urban population was at 17.29% of the overall population. However, this population rose steadily to 27.78% of the total in 2001. The number of 'million-plus cities' increased from 23 to 35 in only 10 years (1991–2001) (Figure 1.4).

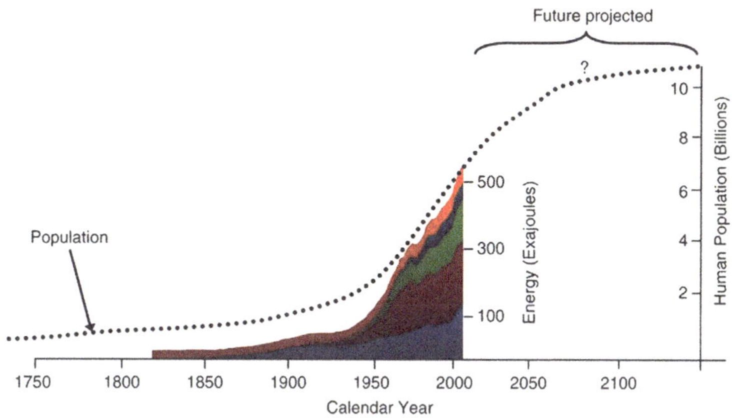

Figure 1.4. World's population change

Source: http://energyskeptic.com/2016/a-strong-case-for-the-anthropocene-no-other-species-has-ever-consumed-as-much-of-earths-resources-so-quickly/human-population-growth/.

Population growth is determined by births and deaths and every country has seen very substantial changes in both: In our overview on how health has changed over the long run you find the data on the dramatic decline of child mortality that has been achieved in all parts of the world. And in our coverage of fertility you find the data and research on how modern socio-economic changes – most importantly structural changes to the economy and a rise of the status and opportunities for women – contributed to a very substantial reduction of the number of children that couples have.

But declining mortality rates and declining fertility rates alone would not explain why the population increases. If they happened at the same time the growth rate of the population would not change in this transition. What is crucial here is the timing at which mortality and fertility changes.

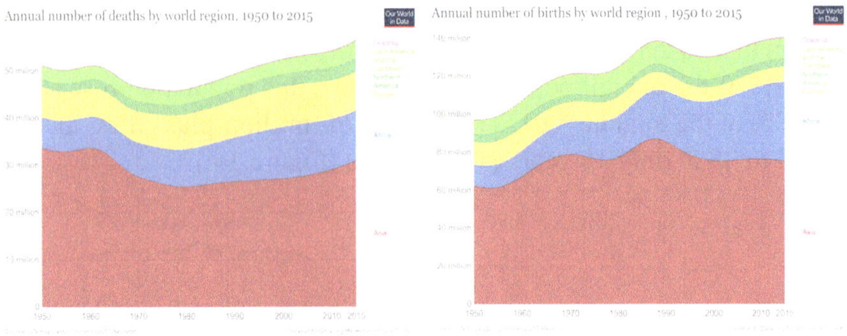

Figure 1.5. Annual Number of Births and Deaths by World Region.

Source: https://ourworldindata.org/grapher/annual-number-of-deaths-by-world-region.

The model that explains why rapid population growth happens is called the 'demographic transition'. It is shown in the schematic figure below. It is a beautifully simple model that describes the observed pattern in countries around the world and is one of the great insights of demography (Figure 1.5).

The demographic transition is a sequence of five stages:

Stage 1: High mortality and high birth rates. In the long time before rapid population growth the birth rate in a population is high, but since the death rate is also high we observe no or only very small population growth. This describes the reality through most of our history. Societies around the world remained in stage 1 for many millennia as the long-run perspective on extremely slow population growth highlighted. At this stage the population pyramid is broad at the base but since the mortality rate is high across all ages – and the risk of death is particularly high for children – the pyramid gets much narrower towards the top.

Stage 2: Mortality falls but birth rates still high. In the second phase the health of the population slowly starts to improve and the death rate starts to fall. Since the health of the population has already improved, but fertility still remains as high as before, this is the stage of the transition at which the size of the population starts to grow rapidly. Historically it is the exceptional time at which the extended family with many (surviving) children is common.

Stage 3: Mortality low and birth rates fall. Later the birth rate starts to fall and consequentially the rate at which the population grows begins to decline as well. Why the fertility rate falls is a question that we answer here. But to summarize the main points: When the mortality of children is not as high as it once was parents adapt to the healthier environment and choose to have fewer children; the economy is undergoing structural changes that makes children less economically valuable; and women are empowered socially and within partnerships and have fewer children than before.

Stage 4: Mortality low and birth rates low. Rapid population growth comes to an end in stage 4 as the birth rate falls to a similar level as the already low mortality rate. The population pyramid is now box shaped; as the mortality rate at young ages is now very low the younger cohorts are now very similar in size and only at an old age the cohorts get smaller very rapidly.

Stage 5: Mortality low and some evidence of rising fertility. The demographic transition describes changes over the course of socio-economic modernization. What happens at a very high level of development is not a question we can answer with certainty since only few societies have reached this stage. But we do have some good evidence – which we review here – that at very high levels of development fertility is rising again. Not to the very high levels of pre-modern times, but to a fertility rate that gets close to 2 children per woman. What level exactly the fertility rate will reach is crucial for the question of what happens to population growth in the long run. If the fertility rate stays below 2 children per woman then we will see a decline of the population size in the long run. If indeed the fertility rate will rise above 2 children per woman we will see a slow long-run increase of the population size.

1.9.1. Sex Ratio

Sex ratio is simply described as the number of females in every 1000 males in a said population. This ratio is a very crucial social indicator used to measure and determine the equality extent between females and males in the population at a specified time. Last 2017, more males (887,972 or 52.2%) were born than females (812,646 or 47.8%) in the Philippines which resulted in a sex ratio of 109 males per 100 females (Figure 1.6).

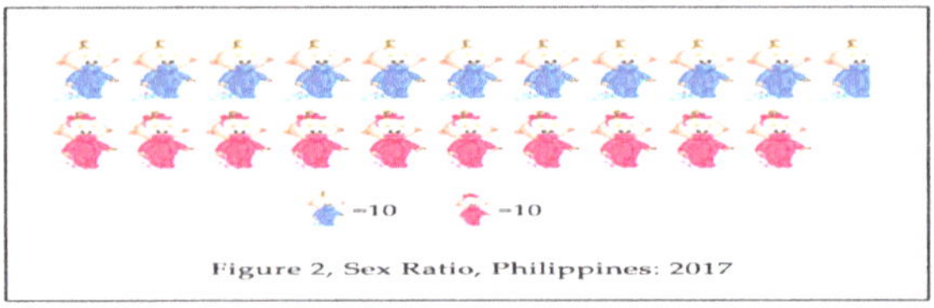

Figure 1.6. Sex ratio.

Source: https://psa.gov.ph/content/births-philippines-2017.

1.9.2. Age Composition

A population's age composition refers to the number of individuals in various age groups in a said nation. Age composition is among the most standard traits of a country's population. To a crucial degree, an individual's age substantially influences their needs, purchases, the activities they are involved in, as well as their performance capacity. Consequently, the percentage and number of a population within the aged, working age and children groups serve as momentous determinants of the overall population's economic and social structure.

1.9.3. Literacy Rates

Literacy is one of the most important qualities of any population. Only an educated and well-informed individual is capable of making intelligent choices and undertake development and research projects. Low literacy levels serve as very serious obstacles for social and more specifically economic improvement and development.

As per the 2001 Census, any individual aged seven and above with reading and writing capabilities in any language whatsoever is treated and considered as literate.

In India, literacy levels have been steadily improving over the years. According to the 2001 Census, the literacy rate was at 64.84%; 53.67% for the female population and 75.26% for the male population. Now, why do differences such as these exist?

1.9.4. Occupational Structures

The economically active percentage of the overall population serves as a very important development index. The occupational structure is defined as the population distribution in accordance with the various types of occupations. There are multiple occupation varieties in any given country. Typically, these occupations are categorized as primary, secondary and tertiary as well.

Primary activities, or rather occupations include animal husbandry, fishing, forestry, agriculture, quarrying, and mining among others. Secondary occupations include production and manufacturing industries, construction, and building works, etc. Tertiary activities, on the other hand, include communications, administration, commerce, and transport among other relative services. The ratio and proportion of individuals working in various occupations vary significantly in developing and developed countries. Developed countries have quite a high number of working folks than the secondary and tertiary sections. On the other hand, developing nations tend to feature higher workforce in primary occupational fields. In India, approximately 64% of the overall population engages in agriculture.

The population proportion that depends on tertiary and secondary sectors is approximately 20 and 13% respectively. Due to the advancements and steady growths witnessed in urbanization and industrialization over the past few years, an occupational shift favoring tertiary and secondary fields has been witnessed.

An urgent call for the residents to each geographical location to nurture their skills and potentials through skill development training so that they could earn more compared on the total reliance on their environment. Government intensive and regular trainings in livelihood for the residents are vital to delimit the consequence.

1.9.5. Health

Another pretty vital constituent of the composition of the population affecting the development process is health. Sustained efforts by various government projects and programs have seen significant improvements and developments in terms of the health of the populations of India. The rates of death have decreased to 8.1% in 2001 from 25% in 1951. Relatively, life expectancy has risen to 64.6 years in the year 2001 from 36.7 years marked in 1951.

This significant improvement is as a result of a number of notable factors including public health enhancement, preventive actions against infectious illnesses as well as adoption and employment of contemporary medical practices and techniques in the diagnosis and healing of various ailments.

Regardless of various achievements and general health improvements, India's health situation still stands as a major concern for its government. Among the issues that rise, concerns do include the inadequate calorie consumption which is way below the recommended levels resulting in malnutrition which afflicts quite a large population percentage. Standard sanitation amenities and pure drinking water are only accessible by a third of the country's rural population. All these short backs need to be handled via an appropriate and relevant population policy.

Based on United Nations Development Programme (2017), more than 1.6 billion people live in fragile settings where protracted crises, combined with weak national capacity to deliver basic health services, present a significant challenge to global health. Good health is essential to sustainable development and the 2030 agenda reflects the complexity and interconnectedness of the two. It takes into account widening economic and social inequalities, rapid urbanization, threats to the climate and the environment, the continuing burden of HIV and other infectious diseases, and emerging challenges such as non-communicable diseases.

1.9.6. Adolescent Population

Indian's population most important feature is the size or rather magnitude of the nation's adolescent population. It covers a fifth of the country's whole population. Typically, adolescents are teens between the ages of 10 years to 19 years. Adolescents are considered as the most significant future resource of any nation. Nutrition requirements and needs of individuals in the said age group are substantially higher than the relative needs of an adult or child for that matter. Keep in mind that stunted growth and deficiency are results of poor nutrition. However, in India, the diet availed to adolescents is deprived of all nutrients. Quite a significant percentage of adolescent girls are suffering from anemia. As of now, their problems haven't received as much attention as they should in the development process. Girls in the adolescent age bracket require sensitization to some of these problems at hand. Awareness on the same can be enhanced through improving education and literacy among the lot.

Population's effects on education and the labor force will influence the opportunities available for the current generation of young people, the largest in history. The demographic transition—the combination of falling mortality rates followed by falling fertility rates—provides countries with a growing number of youth who can be educated and productively employed.

With effective government policies and investments, this transition can create an engine for economic growth.

1.9.7. National Population Policy

In the past 50 years, family planning (FP) programs have been heavily promoted across the developing world as a means to reduce fertility rates and promote economic development. The central assumption behind such programs is that the decline in birth rates during the early stages of demographic transition can promote economic growth, reduce environmental pressures, reduce dependency ratios and strengthen a societies' ability to invest in health and education (Coale, Hoover, and Press 1958). At the micro-level, it has been assumed that a decline in fertility would relieve women of the burden of repeated child-bearing and free up opportunities for them to increase schooling and participate in the labor-force.

Family planning helps people have the desired number of children, which as a result improves the health of mothers and contributes to the nation's social and economic development. In most developing countries, including Ethiopia, it is common practice for women to have too many children, too close to one another. As a consequence, the population size of the country has grown dramatically but economic growth has not kept in parallel with it. Such an unbalanced population size will inevitably have a negative impact on the wellbeing of the nation. Family planning is one of the strategies which are proving to be effective in tackling these problems.

Family planning allows people to attain their desired number of children and determine the spacing of pregnancies. It is achieved through use of contraceptive methods and the treatment of infertility (this fact sheet focuses on contraception). Promotion of family planning – and ensuring access to preferred contraceptive methods for women and couples – is essential to securing the well-being and autonomy of women, while supporting the health and development of communities.

In 1952, the Indian government introduced the 'Comprehensive Family Planning Program.' This is after they realized how far family planning would go in improving and enhancing individual welfare and health as a whole. This great program endorsed and promoted planned and responsible parenthood voluntarily. Years and years of unending planning by the government also resulted in the (NPP, 2000) National Population Policy 2000.

This NPP 2000 provides the required framework to impart compulsory and free education to children less than 15 years of age. This, in turn, reduced infant death rates to less than 30 mortalities in every 1000 births. It also saw to global immunization of kids against any vaccine-preventable ailments and promoted girls' late marriage thus ensuring family welfare became the center of focus for every other family.

1.9.8. The National Population Policy and Adolescents

The National Population Policy 2000 identified and recognized adolescents (aged 10 to 19) as one of the key population segments requiring much attention. Apart from the basic call for their nutritional needs, this policy also emphasized and spearheaded various paramount needs for this age group including protection against sexually transmitted diseases and unplanned or rather unwanted pregnancies. It also laid great emphasis on:

- Delayed, or rather late marriages and bearing of children;
- Adolescent education on the risks one is opened to when they indulge in unprotected sex;
- Ensuring that contraceptive services are affordable and easily accessible by the same;
- Enhancing their health by providing nutritional services and food supplements;
- Adopting new, and strengthening the available legal measures preventing the marriage of children.

With a description of the socio-cultural milieu of adolescents in South and Southeast Asia, their knowledge and use of contraceptives, the myriad barriers to access, and the many innovative programs to broaden contraceptive availability. The reproductive health needs of adolescents in poor countries cannot be solved by merely supplying them with contraceptives--these needs can only be fully addressed in the context of gender equality, poverty alleviation and the conviction that investing in the reproductive health of adolescents is a most urgent priority. Investing in the reproductive health of adolescents will have an impact not only on birth and abortion rates, maternal health, and the spread of STI/HIV but also on the demographics and economic development of the region--and beyond.

Half of all pregnancies in the predominantly Catholic Philippines are unintended, according to a recent study by the Guttmacher Institute, a US-based think tank that promotes reproductive health. Of those unintended pregnancies, 90 percent are due to a lack of modern methods of contraception. Unlike in some other developing nations, the Philippines' government has not provided free contraception to those who need it most.

The country begins to roll out free contraception under its new reproductive health law. The groundbreaking legislation requires government health centers to distribute free contraceptives, including birth control pills and condoms, no matter the beliefs of the politicians in power. The law also aims to educate Filipino youth about family planning by integrating sex education into school curricula. The path to free family planning services for the poor wasn't an easy one. Fierce opposition from influential Catholic Church leaders, who have long played a major role in politics, resulted in a bitter 14-year battle in Congress.

The youths have a strong need to receive contraceptives most especially in rural communities to foster awareness to avoid to unwanted pregnancy. According to Zosa-Feranil (2003), there are many operational barriers to sexual and reproductive health care and supplies exist for Filipino adolescents, including cultural stigma against (unmarried) youth using contraceptives, negative attitudes on the part of health service providers, pressure from the Church, and lack of adequate supplies at the local health system level. Marital status is used as basis for providing reproductive health services: Because of the sensitive nature of Adolescent Reproductive Health issues in general, marital status instead of age has generally been the major factor determining how sexual and reproductive health programs are organized and provided. As the head of a prominent NGO working with youth and adolescents explained, a married adolescent would not be treated as a teenager but as an adult; access to contraceptive services is considered appropriate for married women. Politically, age is not really the issue in this country but premarital sex. However, despite resistance, there is clamor to focus attention on younger adolescents and to adopt the exact WHO definition of adolescence, which extends between the ages of 10 and 19.

As a fact, people or the citizens of a country are its most valuable asset or resource. Therefore, ensuring that the lot is well-educated will undoubtedly provide and endorse potential power.

1.10. MODERN SPATIAL DEMOGRAPHY AND POPULATION GEOGRAPHY

When tackling, or rather handling population from a geographic perspective, there are three main or primary themes:

- Place-based population;

- Spatial variation; and
- Human-environment interaction.

Geographers are more focused on spatial variation which seeks to understand and comprehend why and how things vary and differ from one place to another across various locations. They seek to understand how these spatial patterns change or evolve. The focal point of geographical studies is the interaction between the environment and human beings; ways in which the physical surrounding and environment impact the society along with how the society adapt to or change their environment as time goes by. The last theme looks to understand the well-being of populations based on places. It also covers the characteristics of these regions or places. Population geography covers every single one of the three themes mentioned above in a bid to comprehensively understand the study of population and places.

1.10.1. Spatial Thinking and Geographic Concepts

The most basic or standard concept of geography is a location- what or where exactly is a place? As much as this is true, understand that even this very simple question proves very difficult when other spatial thinking aspects are not employed. Site and situation are among the most fundamental concepts of geography. A site can be simply described as the traits or rather characteristics of a place while situations are how different places are connected by either human activities such as communication and transport or natural undertakings. A geographer's basic analysis of the situation and site of a given place goes way beyond on the mere recognition and appreciation of its boundaries. According to Gershmel (2008), for any population geographer, 'place' is considered as a very complex term that raises, or rather invokes several other spatial thinking dimensions including the concepts of analog, aura, comparison, connection, distribution, gradient, influence, pattern, region, relationships, proximity, mobility, interaction, hierarchy, exceptions, distance decay, diffusion, etc. Tobler (1970) also states that geography's first law is that every single thing is related to every other thing in one way or another. However, closer things tend to be much more related as compared to things situated further apart.

Usually, geographers are known to seek answers to questions connecting demographic occurrences over space and time. Folks and places are interwoven intricately, both temporally and spatially. It is not unusual for geographers to seek answers on the differences or similarities of places. How does a specific place compare to another, or a country, or even to the universe at large? How does a specified attribute influence or affect the areas surrounding it? Do regional distinctions exist between demographic events? What is the essence of inter-places transitions? Are the landscape gradients abrupt or gradual? Do any spatial designs such as clustering and imbalances among any distinctive patterns exist or do the phenomena occur randomly? Is there any evidence of spatial correlations or associations? Are there any locations or places that defy the typical rules? Also, how do things proliferate across space through time? What barriers or rather avenues seem to either enhance or hinder diffusion? According to Gershmel (2008), these are the kind of queries posed by spatially-thinking among geographers. In short, properly understanding and comprehending the concept of spatial thinking when handling demographic sightings is the bosom of population geography.

The magnitude of scale is highly appreciated by population geographers. The analysis scale plays a very vital role in determining geographic representation. You rarely come across a question of the right kind of scale rather than the appropriateness of the said scale. The more important question is; exactly how far does this scale analysis goes in governing the outcome of a study? There are a number of studies and research that are best conducted with a global point of view. Such include any studies related to pandemics, global climate, as well as human rights. Whereas, there are other studies that are emphatically concerns of household families. An example of such is research on completed fertility where spatial aggregation elevates as the analysis scale progresses along the hierarchy of geography from an individual, to people, to families, along with neighborhoods, towns, or cities, all the way to the universal context. Also, keep in mind that as noted, there are several other analysis units that don't perfectly fit in a geographic hierarchy. These groups kind-of surpass traditional scales includes mankind congregations, stakeholders, and unions of countries among several others.

The Modifiable Area Unit Problem (MAUP) can be simply described as the unpredictability amalgamated with spatial aggregation. Initially defined by Openshaw in 1984, all areal measurements employed in most geographical research and studies tend to be modifiable, arbitrary, and susceptible to the aggregating researcher's notion. Geographers take various approaches when contending with MAUP. A number of studies undertake similar analyses in varying spatial aggregation levels in order to determine their outcome's relative sensitivity. Others, on the other hand, deliberately jump scales in a bid to generate alternative knowledge. Just like any other spatial science, in population geography, MAUP twirled around the empirical defiance of data integration across various geographic scales.

2

PATTERNS OF POPULATION COMPOSITION

CONTENTS

2.1. INTRODUCTION

A population is made up of different components. This chapter was prepared with one aim of presenting different patterns of population composition as discussed below.

2.2. POPULATION AGE STRUCTURE

Population age structure is the percentage of people in a population that falls in various ages, classes, and sexes. We can also define the population age structure as the number of males and females present in various age groups within a particular population. Population structure shows the divisions of a population on the bases of sex and age. The dynamics of a population are dependent on the age structure. The age distribution of the members of a population changes its age structure. Population age structure is represented in graphs known as population pyramids.

2.2.1. Population Pyramids

Population pyramids are a type of graph made up of stacked bars. The bars are horizontal. The proportion of the population is shown on the X-axis. The Y-axis depicts age-groups within the population. Usually, population pyramids do not indicate the actual figure of the members of a population. Rather, it shows the percentages of the population in the various age groups. Measures of central tendency must be put into use to enable accurate reading of population pyramids. A population pyramid helps make several deductions about a population such as birth rates, mortality rates, dependency ratios, and life expectancies. The dependency ratios of a population are established by determining the economically active members of a population and gauging the proportion against the less productive and unproductive members (Figure 2.1).

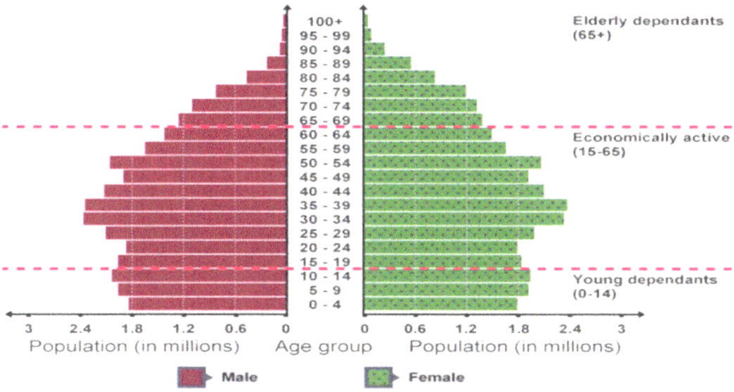

Figure 2.1. Dependency ratios of a population.

Source: https://www.bbc.co.uk/bitesize/guides/ztr2w6f/revision/1.

Population pyramids vary in shape depending on the age distribution of the represented population. A triangular-shaped population pyramid represents a population that consists of a higher proportion of younger people. As such, a population is fast growing. A population pyramid that is column-shaped on the other hand, represents a population made up of a higher proportion of older people. Such a population is slow-growing (Rosset, 2017).

2.2.2. Types of Population Pyramids

Population pyramids come in three different types.

1. **Expansive Pyramids:** The first type is an expansive population pyramid. Expansive pyramids are used to represent populations whose larger percentage is made up of young people. There are two major characteristics that populations with a higher proportion of younger people portray. First, their fertility rates are very high. The life expectancies of this type of population are also significantly low. An expansive population pyramid is triangular in shape. Expansive population pyramids are commonly found in most developing countries. An illustration of an expansive population pyramid is given in Figure 2.2.

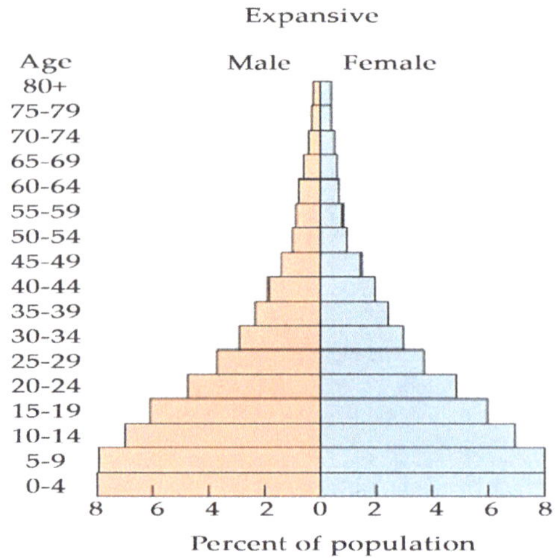

Figure 2.2. An expansive population pyramid.

Source: https://populationeducation.org/what-are-different-types-population-pyramids/.

2. **Constrictive Population Pyramid:** The second type is constrictive population pyramid. The bottom of this type of population pyramid is small, hence the term was constricted. The population represented in this type of pyramid is one composed of many older than younger people. This type of population records low birth rates as a result of less productive people. The dependency rate in this type of population pyramid is high since the economically productive population is low. This is a population pyramid typical in countries that are very developed. The higher population is usually educated and knowledgeable about birth control methods. Health care services are also good. Additionally, factors that affect the environment negatively are few. The use of birth control reduces the number of children born into the population. On the other hand, safe environmental practices and well-developed health care services reduce the number of deaths among members of the population (Rosset, 2017). An illustration of a constrictive population pyramid is given in Figure 2.3.

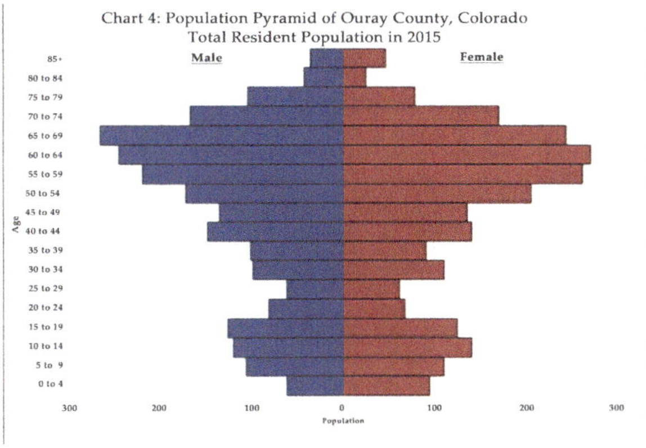

Figure 2.3. A constrictive population pyramid.

Source: https://www.census.gov/newsroom/blogs/random-samplings/2016/06/americas-age-profile-told-through-population-pyramids.html.

3. **Stationary Population Pyramid:** A stationary population pyramid is the last type these graphs. This type of pyramid represents a population that has almost equal proportions of people in all the age classes. The type of population captured in this type of pyramid records equal birth and mortality rates. Usually, this type of a population does not decline nor increase. An illustration of a stationary pyramid is given in Figure 2.4.

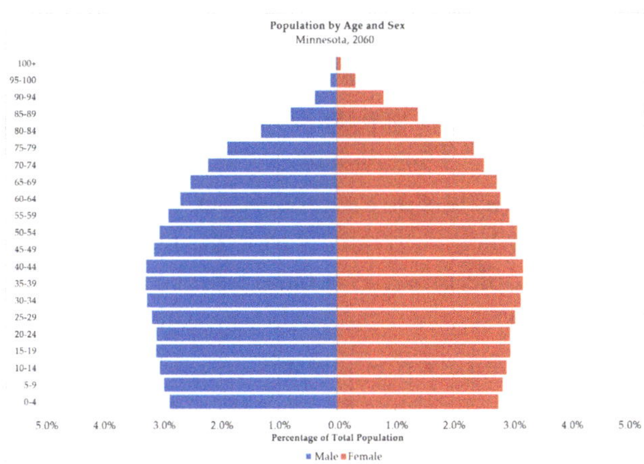

Figure 2.4. A stationary population pyramid.

Source: https://www.yourvoteyourvoicemn.org/future/developing-trends/community-demographics/minnesota-demographic-forecast.

Population pyramids can be used to determine various characteristics of a population. The characteristics include the effects that activities such as migration, war, scarcity of food and sicknesses bring on a population. You can also deduce information about the life expectancy of a population from the pyramids.

2.2.3. Stages of Population Pyramids

Population pyramids have five stages. All the stages have different characteristics.

1. **First Stage:** The first stage is characterized by reduced life expectancy and increased birth rates. The mortality rates in this stage are high. The people that attain old age in this stage are very few. The first stage population pyramid has a wide base and a concave profile.
2. **Second Stage:** The second stage of population pyramid records life expectancy that is low. Birth rates in this stage are higher. The second stage, therefore, has more children than adults. The dependency rates in this stage are very high since the higher proportion of the population is not economically productive. The population pyramid in the second stage has a wide base. The profile is pyramidal. The life expectancy in this stage is high.
3. **Third Stage:** The population in the third stage has equal birth and mortality rates. Additionally, life expectancy in this stage is high. Dependency rate is also low and the population in this stage is stable. The population pyramids in the third stage has a narrowing base. They also have a space rocket profile.
4. **Fourth Stage:** It is on the other hand, begins to exhibit itself in a population when birth rates start to decline making the proportion of older people more. Population pyramids in this stage have a narrowing base. Their profile is vase-shaped.
5. **Fifth Stage:** Lastly, this is an advancement of the fourth stage. It is characterized with reduced life expectancies resulting from more deaths in the population. The base of the population pyramids is very small.

2.3. POPULATION ECONOMIC COMPOSITION

Population economic composition is a division of the population on the basis of their economic state. Members of a population live within different economic situations. Some are economically stable while others live below the poverty level. There are various reasons why there are different economic classes in a population. The reasons can be social, cultural or environmental (Holdsworth et al., 2013).

2.3.1. Social Factors

There are several social factors that influence the economic class in which members of a population will be. Some of these social factors include:

1. **Education:** Not every member of a population manages to get an education. Also, some go to school but fail to achieve good grades. On the other hand, a proportion of the population is smart but is unable to achieve high levels of education to various circumstances. Another section of the proportion of population manages to achieve very high levels of education.

In the professional job market, staffs are employed to positions that correspond to their academic levels. It is worth noting that the higher the position, the better the pay. A person with little or no education can only qualify for a job position in a low level. A job position in a low level only attracts a small pay. On the other hand, a person that has a high education level secures a job in a high position with a better pay.

The proportions of the population that do not get a chance to achieve a high level of education are thus less likely to get high job positions. They are therefore less likely to get good pays. Contrariwise, the well-educated proportions of a population are more likely to get employed into higher positions therefore getting good pays.

The better the pay, the better the living conditions. Also, the better the pay, the higher the economic class. The less educated members of a population live in a lower economic class.

The proportions of the population keep changing. The more the people getting a high education, the higher the number of members in the higher economic class gets. If more people are unable to get a high education, the more the proportion of the population in the lower economic class gets.

Educational opportunities are vital to ensure sustainability of economic resources. Access to any forms of education can help in developing once skills and competencies necessary to earn a living. In most developing countries, the rural areas do suffer from several basic problems and among them rural livelihoods have emerged as a crucial issue for the policy. Despite the development efforts, the majority of the rural population still endures absolute poverty (Singh, 2010). Poverty is strongly associated with low levels of education and skills.

Educational opportunities heighten the consequences of rapid population growth in different locations and levels. A strong government intervention giving a profound and inclusive access or opportunities to education to lessen population pressures affecting the communities is necessary. Policymakers should analyze and help develop programs that balance social, economic and environmental goals with demographic shifts.

2.3.2. Cultural Factors

Cultural factors revolve around the norms, traditions, and beliefs of the members of a population. Some cultural factors that influence the economic composition of a population include:

1. **Early Marriages:** In a population where the members practice early marriages, the economical state is bound to be affected. The members that get involved in early marriages are denied the chance they get a good education. Without a good education, they are less likely to secure employments that attract good pay. They are therefore more likely to belong to the lower economic class.

The proportion of the population that get an education rather than get married at an early age stand a chance to secure employment that will secure a sound and abundant life. They are therefore more likely to belong to the higher economic class (Holdsworth et al., 2013).

2. **Inheritance:** It refers to properties that are carried forward from one generation to another. A family that is wealthy is more likely to leave their children enough wealth. On the other hand, children from a family that does not have much to leave them are forced to start gathering their wealth from scratch. The proportions that inherit much wealth are more likely to gather more wealth sooner than those that inherited little or no wealth.

People that come from wealthy families are more likely to belong to the higher economic class than those whose families are not well up.

3. **Cultural-Economic Activities:** These refer to the undertakings that members of a population do to earn a livelihood. Some cultures carry out activities that earn them a lot of money. Others engage in activities that earn them just a little. The proportions of the population whose cultural-economic activities earn good money are more likely to belong to the higher economic class.

On the other hand, the proportion of the population that carry out cultural-economic activities that earn them a little money are more likely to belong to the lower economic class.

2.3.3. Environmental Factors

Environmental factors also play a part in determining the economic class that members of a population will belong to. Such factors include:

1. **Topography:** The topography of a place determines the economic activities that the inhabitants of an area will carry out. The activities carried out in a hilly area are different from the activities carried out in a flat topography. For instance, highlands support the rearing of dairy cattle and crop farming. The activity earns the inhabitants some money (Figure 2.5)

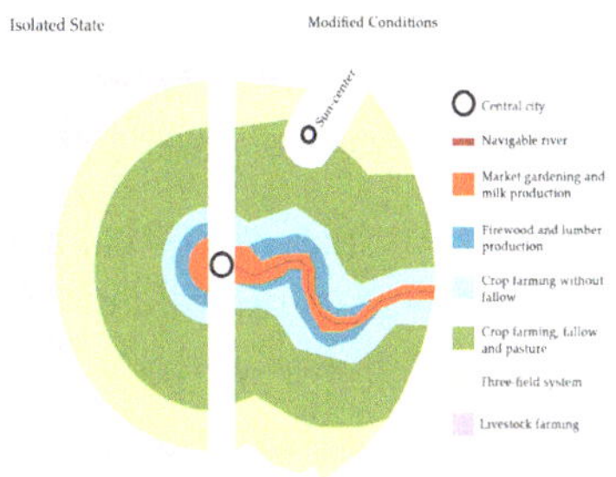

Figure 2.5. Von Thunen's Regional Land Use Model.

The inhabitants of lowland, on the other hand, are likely to buy food and food products from the population living in the highland. The economic class of the population in the highland is, therefore, more likely to belong to a higher economic class that the population living in the lowlands (Gould, Population, and Development, 2015).

2. **Temperature:** Temperatures of an area determine the activities that the population living in the area will undertake. For instance,

the crops grown in hot areas are different from the crops grown in cold areas. The crops that survive in cold temperatures include tea, apples, and pears. The crops grown in hot areas, on the other hand, include rice and coconuts. Depending on the market prices of the crops grown, the inhabitants of climatic region may earn more from the sale of the crops they grow (Figure 2.6).

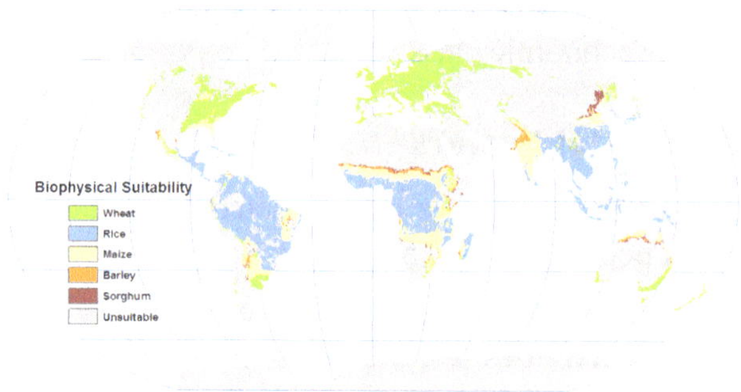

Figure 2.6 World's Biophysical Suitability.

Source: UNEP (2010): The GEO Data Portal, as compiled from FAO, TERRA-STAT I Global GIS Databases Poverty and Food Insecurity Mapping Project. United Nations Environment Programme. http://geodata.grid.unep.ch.

The population growing crops that are sold at a higher price are therefore more likely to belong to the higher economic class. On the other hand, the proportion of the population that major with the farming of crops that go at a lower price are more likely to belong to a lower economic class.

The temperatures of an area also determine the type of livestock that the inhabitants rear. The proportion of the population that live in a region whose temperatures allow for the rearing of animals that earn the farmers good money from the sale of their products.

2.4. POPULATION ETHNIC COMPOSITION

Population ethnic composition refers to the proportion of the various ethnic groups making up a population. Ethnic groups, on the other hand, are made up of people with a common culture, ancestry, and live in a common region (Figure 2.7).

Figure 2.7. Populations can be made up of one or more ethnic groups.

Source: http://www.digiverse.ca/ethnic-market-research/.

Some ethnic groups may be more within a population than others. The proportion of the various ethnic groups is influenced by several factors.

2.4.1. Factors Influencing Population Ethnic Composition

The high or low proportions of ethnic groups within a population are result of various factors. Such factors can be majorly divided into groups:

1. **Social Factors:** These are the things that people do as a way of life. Such factors affect the composition of various ethnic groups within a population. Social factors include:

- **Intermarriages:** When people live together, they are bound to associate in a number of ways. Marriage is among the associations. Intermarriage is the association in marriage between members of different ethnic groups. When a person gets married into an ethnic group different from their own, they represent their ethnicity in the population. When many people with a common ethnicity marry into a similar ethnic group different from their own, then the proportion of their ethnic group grows. If for instance there were no members of their ethnic group in that population, then through marriage, their ethnic group becomes part of that population. The more the people from this ethnic group that marries into the population, the higher the proportion of the group. Contrariwise, their ethnic proportion in the populations they are marrying from reduces.

- **War:** It results from various commotions within a population. When war breaks, the peace of an area is disrupted. Deaths may also result when there is war. War results from various causes such as political instabilities, land disputes among others. People run for their lives during the war. They increase the proportion of their ethnic composition in the place they move to. Their movement also reduces their ethnic compositions in the places they migrate from.
- **Proximity to Social Amenities:** These are the facilities that offer essential services that people need to make living better and easier. It may be difficult and costly to live in places far from social amenities. Therefore, people tend to prefer living in areas where these facilities are available. Some people move from the places they leave to places where such facilities are easily accessible. Their movement may have an effect on the population's ethnic composition. They may cause an increase in the ethnic composition of their respective ethnic groups in the areas they move to. However, the places they move from experience a decline in the ethnic composition in the ethnic groups to which they belong.

2. **Economic Factors:** These are aspects that involve the peoples' source of livelihood. These aspects also influence ethnic compositions in a population. Economic factors that influence ethnic composition include:

- **Employment:** One of the sources of livelihood today is employment. When a person is looking for a job, they may not have the liberty to choose the location they prefer. They may, therefore, be assigned to operate from a place away from their home. The population of the place they are posted may be void of members from their ethnic groups. When they move to the area, they become a part of the population in that place. They introduce their ethnic group into the population. If other people from their ethnic group are assigned to work from this area, their ethnic composition rises.
- **Business Opportunities:** These may lead people to move from their place to another. It is usual for people to be drawn to areas where they are likely to make money. Some places have better business opportunities than others. People may be attracted to an

area that is made up of ethnic groups other than their own. When several people from a common ethnic group move to another area to do business, they introduce or increase the composition of their ethnic group in that population.

3. **Environmental Factors:** These include the climatic conditions and the topography of an area. These factors also influence the population ethnic composition. Environmental factors include:

- **Drought:** This may lead to scarcity of food and water. Food and water are basic human needs. It is impossible for humans to survive without food and water. When famine befalls an area, its population may be forced to move in search of food and water. In their quest for food, they may settle in other areas. They become part of the population of the area they move to. The composition of their ethnic group increases in the area. The proportion of their ethnic group goes higher if the members of their group that moves to the area are many.
- **Floods:** It results from too much rain. Lowlands also experience floods when there is too much rain on the highlands surrounding them. Floods are hazardous and are capable of killing members of a population. People may migrate from their places to seek refuge in other areas not affected by floods. The ethnic composition of their ethnic group into the population increases when they settle in that area.
- **Landslides:** This may result from too much rain or movement of the lithosphere. Landslides may bury property and lead to the deaths of people. When they happen, the people living in that area are forced to move to other areas to find safety. The areas they move to experience increased ethnic composition in the ethnic groups they belong to. On the other hand, the areas they move from experience a decline in proportion in the ethnic group they belong to.
- **Volcanic Eruptions:** Volcanic mountains erupt from time to time. Volcanic eruptions are hazardous to the population living close by. They may lead to death of a section of the population. The ethnic compositions of the ethnic group whose members die from the eruption suffer a decline. On the other hand, the survivors may settle in a further place leading to a further decline in their ethnic composition.

2.5. POPULATION FAMILIES AND HOUSEHOLD

Every population is made up of families and households (Figure 2.8).

Figure 2.8. Families and households are the most basic unit of any population.

Source: http://www.cursoer.com/en/families-and-households/.

The two differ in meaning and composition. A family is made up of three or fewer people that are related by blood. Adopted children are also part of a family. On the other hand, households are made up of three or fewer people that share a house, home or apartment but are not blood-related (Hareven and Vinovskis, 2015).

2.5.1. Types of Families Found in a Population

1. **Nuclear Family:** It is one of the most common types of family. It comprises of married partners and the children they have given birth to or adopted. You can find four types of nuclear families.

- **Single Parent Family:** This type of family has only one parent. The children form a single parent family depend on the one parent present to provide for their needs. There are several reasons that result in single parenthood such as the death of one partner, divorce among other reasons. The proportion of single-parent families in a population can change based on these reasons. For instance, when one partner in a marriage dies, that family automatically becomes a single-parent family. On the other hand, if a parent gave birth to a child out of wedlock and later remarries, that family ceases being a single-parent family. This reduces the proportion of single-parent families in a population.

- **Two-Parent Families:** It is another type of nuclear family. It is made up of two partners that are in a marriage and the children they have either given birth to or adopted. The proportion of two-parent families can increase in a population when more people get married and have children. The percentage of two-parent families can also reduce when the marriages are broken through divorce or death. When marriages are broken, the families change from two-parent to single-parent families. Also, one partner in the marriage can marry another partner changing the family type from two-parent to polygamous or polyandrous.
- **Polygamous Families:** It is one where the man involved marries more than one wife. The children he bears with his wives are his dependents. Polygamous families are more in a population where the culture allows for a man to marry more than one wife. In such a population, the percentage of polygamous families may be higher than other types of families. Polygamous families increase in proportion when a previously monogamous husband marries another wife. The proportion can reduce when the husband dies or divorces one or several of his wives. When the man dies leaving his wives behind, their families become single-parent families. It is worth noting that when a polygamous man divorces one or several of his wives and still remains with more than one wife, the family still remain polygamous. However, the family of the divorced wife becomes a single-parent family.
- **Polyandrous Family:** Most cultures in the world are male-dominated. It is, therefore, less common to find women being the head of their families. Polyandry is said to have taken place when a woman marries more than one husband at the same time. Polyandrous families are more in proportion where the culture of the land allows for the same. They are also few where the culture does not allow a woman to marry more than one man. Polyandrous families decline in percentage when the woman in the marriage

dies. The proportion also goes down when the woman breaks her marriage with one or some of the men to whom she is married. The proportion goes high when more women within a population marry more than one man.

2. **Extended Families:** These are believed to be the most common type of marriage. They are composed of three or fewer generations from common ancestry. The members that make up this type of family are marriage partners, their children, and their grandchildren. Nuclear families are part of an extended family. Nuclear families also grow to become extended families when they get to their third generation.
3. **Blended Families:** It is formed when two divorcees get married. Two people in a widowed relationship also form a blended family when they come together in marriage. A divorcee and a person in a widowed relationship can also form a blended family when they get married. It should be noted however that a blended family is only formed when both or one of the parties in this kind of marriage brings children into the marriage. These children must be from their past relationships. They may then have other children in their current relationship.

The proportion of blended families in a population increase when more divorced or widowed persons get married. On the other hand, when this marriage breaks, the number of blended families in a population is bound to reduce (Hareven and Vinovskis, 2015).

4. **Family by Choice:** It is the newest type of family. This type of population was defined by the LGBT community as the type of family that one chooses to be part of. The people you choose to be your family members are not your blood relatives. One, however, leaves their real family and becomes a part of another family. The people they consider their family may be from diverse families but share a common interests.

2.5.2. Types of Households

The population is also made up of various households. Some households may be more in proportion in a population than others. The variance is a result of several reasons. Below are the types of households that you can find in a population:

1. **Census Households:** A census household is made up of at least one family that is enumerated as one during a census.
2. **Non-Census Household:** This is a type of household that carries a single person who lives in a particular home alone. Several people living together in the same household but cannot be enumerated as one family, also make up a non-census household.

2.6. POPULATION LANGUAGE

Population language refers to the division of population on bases of the languages they use to communicate. Language can be acquired or learned. Acquired language is known as first language. On the other hand, all the learned languages are known as second languages. Some languages have more speakers in a population than others. Historically, particular languages were used in distinct areas. The users of one language were found in a common region. However, today, users of one language can be found in different places at a go. There are several reasons why the proportions of speakers of some languages are higher than others within a single population. Similarly, several factors influence the number of users of different languages in a population. Such factors may be social, economic, and cultural (Holdsworth et al., 2013).

2.6.1. Social Factors That Influence the Proportion of Users of Different Languages in a Population

Social factors refer to the aspects of the people's way of life that influence how many users the languages used have. It is worth noting that some languages tend to attract more users than others. Some of the social factors that have an effect on the proportion of language speakers include:

1. **Education:** It is a social aspect that affects many areas in the human life. Population language is no exception. When people go through school, they are exposed to various languages. Some of the languages are taught as part of the curriculum. They also learn other languages from the people they interact within the course of schooling. People may adopt the languages they learn and start using them in their conversation. Once a student decides or is obliged to use a particular language, they increase the proportion of users of that language in the population.

2. **Historical Factors:** These in language involve all the aspects revolving around history and amount of affection a language used in a population. All language users are believed to have cropped form a particular place. History suggests that the users of one language have a common origin. Therefore, it is likely to find a higher proportion of users of a particular language around the place where the languages users originated or settled. It is believed that some language communities moved from their places of origin and settled in a different area. These places usually experience a higher population of users of that language than the percentage of users of other languages.
3. **Intermarriages:** Since a long time ago, users of different languages have been intermarrying. When the users of one language marry into another language, they may need to learn that language. They may need to use this language to enable them to effectively communicate with the speakers of this language. They may even start using that language as their mode of communication. They may thus stop using their native language altogether and adopt this new language. That way, the languages they adopt experience a higher proportion of users. On the other hand, the languages they neglect experience less proportion of users (Holdsworth et al., 2013).

However, it should be noted that it is unlikely for an adult to forget their native language even if they do not use if often. This is especially unlikely when they have used that language for a significant period. However, the new languages they learn may affect the usage of their first or other languages they may have learned. Note that whether they forget the other language or not, they become members of the language they use to communicate often.

4. **Religion:** It is the belief in a power or being usually considered to be supreme. Some religions require their followers to learn a certain language to enable them understand the religion. Some religious books are also written using a particular language. Non-users of the language may find it difficult to understand the content in these sacred religious books. They may therefore resort to learn the language to make understanding of the religion easier. The more the people that learn a particular language, the higher the proportion of the users of that language.

If the people that learn the language of their religion end up making of their main language of communication, they may end up neglecting their former language. Consequently, their former languages experience a decline in the proportion of users (Holdsworth et al., 2013).

2.6.2. Economic Factors That Influence the Proportion of Users of Different Languages in a Population

Economic factors refer to the aspects related to finances that people within a population partake in. These aspects include:

1. **Attracting Clients:** When a person establishes a business in an area where most of their potential clients use different languages from their own, they may experience difficulties doing business. It may be hard for them to convince clients to buy from them. They may even lose clients to other business owners who can communicate using the language of the clients. Such business owners may be forced to learn the language of their clients to be in a position to draw more clients their way. Learning the language breaks the language barrier between them and their clients. They may then resort into using this language as their mode of communication. Consequently, the proportion of users of this language in the population increases. Also, the percentage of users of their former language decreases.
2. Learning Business Strategies: Business strategies involve the things that a business needs to implement to make more profits. When one establishes a business in a different area from their native land, they may encounter different people that use different languages. The secret of thriving in business in this area may be hidden in the language they use. Remember, different people from different areas and cultures may prefer different types of products and services. The best way to learn the needs of the community is being associated with them. Association requires using a common language to share ideas. Therefore, this business person may be obliged to learn the language used by their target clients. This way, they increase the proportion of the users of this language and reduce the users of their former language within the population.

2.7. POPULATION MARITAL STATUS

Population marital status is the categorization of population using their respective marital status. Marital status can be described as the state of population members with regards to marriage. Marital status can also be referred to as civil status (Figure 2.9).

Figure 2.9. Different members of a population belong to different marital statuses.

Source: https://www.buddyloans.com/news/uncategorized/does-your-marital-status-affect-your-car-insurance-premiums-112099/.

2.7.1. The Various Types of Marital Status

1. **Single Marital Status:** A person that has never been involved in any legal marriage is said to be single. The status changes as soon as they get married. From there, the status could change to any marital status based on situations.
2. **Married Marital Status:** As soon as a person is part of a legal or recognized marriage union, their civil status is married. The person remains married as long as they remain in that union. Their marital status changes when they break out of that marriage. Also, intentional or accidental termination of that marriage due to various reasons automatically changes a person marital status from married to divorced, separated, or widowed.

- **Separated Marital Status:** A person gains the separated marital status when they break of their marriage. In this status, however, the people involved are still legally married. Separation is the initial phase of divorce. People stay separated until they finalize their divorce process. However, there are marriage partners that get back together after separation. In this case, they gain back their married marital status.

- **Divorced Marital Status:** Divorce happens when people that were linked together in a marriage fall out. The parties involved undergo a legal process to be pronounced divorced. After a divorce, either of the parties is free to remarry according to the law.
- **Widowed Marital Status:** Death is inevitable. People die due to various reasons. The reasons vary from person to person. When one party in a marriage dies, the other party marital status changes from married to widowed. It is permissible per the law for the widowed party to remarry.

Marital statuses vary in proportion within a population. The proportion of some marital statuses within a proportion may be higher than others. Some of the factors that cause variance in the percentages of members in the different marital statuses are affected by social factors.

2.7.2. Social Factors That Influence the Proportion of Marital Statuses in a Population

Social factors refer to the aspects that revolve around the way of life among the members of a population. Some social factors that influence the percentage of the various types of marital status in any population are:

1. **Education:** It is among the social factors that influence the number of members of a population that belong to a particular type of marital status. Education determines how old the members of a population will be when they get married. Usually, people start undergoing school at a young age. Some people get to high levels of education. Others only get some basic education, while others go beyond basic education but do not manage to get to high education levels. The rates at which members of a population grasp information taught to them through education determine how far they pursue education. Also, the financial capabilities of their parents or guardians also determine the level of education that people get to (Bogue et al., 2010).

The time that people take in school, determine how soon they get married. The proportions of the population that spend just a little time in school are more likely to get married sooner. Once they get married, the proportions of the population under the category of single marital status reduce. On the other hand, the proportion of the population that is married increases.

Contrariwise, the proportion of the population that takes a longer time in school is more likely to get married at an old age. Remember, schooling takes a lot of a person's time and attention which may lead this proportion to lack sufficient time to engage in marriage. The longer this proportion remains unmarried, the longer the single marital status proportion in the population gets higher. Once they get married, however, the proportion of married marital status increases in the population.

2. **Alcoholism and Drug Abuse:** Alcoholism and drug abuse have long been a menace in the society. People that engage in alcoholism and drug abuse are less likely to engage in marriage. If members of a population indulge in alcoholism and drug abuse before they get married, they reduce their chances of getting married. Remember, everyone looks forward to a happy marriage. It is, however, unlikely to have a happy marriage when your partner is an alcoholic or abuses drugs. The proportions of single members of a population that abuses drugs or is in alcoholism are thus less likely to change their marital status. This makes the single marital status proportion in a population higher. On the other hand, people that engage in drug abuse and alcoholism when already in marriage are less likely to sustain a marriage. It is worth noting that alcoholism and drug abuse attract many negative activities, effects, and impacts such as violence. Many parties that are married to embers of a population that engage in alcoholism and drug abuse do not enjoy happy marriages. A proportion of these members resort to walk out of a long time commitment in relationship. Once they walk out, the proportion of married members of a population reduces. Also, the percentage of separated or divorced members increases (Bogue et al., 2010).
3. **Diseases:** These are natural ailments that befall people making their bodies weak. Diseases also often lead to deaths. Some diseases are deadly and fierce. When members of a population contract dangerous diseases when still single, they are unlikely to attract marriage partners. They are therefore more likely to

remain single. This increases the proportion of members of the population that are single.

Additionally, there are instances where diseases cause deaths. When that happens, they influence the marital status of the members of a population. The decline on the proportion of the population where the deceased persons belong is observed. On the other hand, if the deceased persons were married, their spouses change their marital statuses from married to widow after they die.

2.8. POPULATION NATIONALITY

Nationality refers to the country where one comes from or originates. It is the legal or formal right of an individual to belong to a certain country. A population with the same nationality is bound together by the same race, religion, language, or traditions. Different people with different nationalities can be found in a single population. However, some nationalities can have a higher proportion of members than others in the same population. The differences in numbers can be attributed to several reasons. It is worth noting, that one can acquire the nationality of a country in several ways. The different ways of acquiring nationality include birth, naturalization, resumption, by subjugation and through cession (Boll, 2007).

2.8.1. By Birth

Majority of populations acquire nationality just by being in that country. It can be according to jus soli (right of soil) whereby the nationality of the child is determined by the place of birth or according to jus sanguinis (right of blood) whereby nationality comes about if one or both of the parents are citizens of that nation.

The more the number of children bore to a particular nationality, the higher the proportion of that nationality in that population.

2.8.2. Naturalization

This is where the citizen of a foreign nation is living in another nation and acquires citizenship of that nation. The rules of naturalization vary from country to country but commonly include the promise to obey and uphold the country's laws. Naturalization takes places in various ways:

1. **Marriage:** This is when a foreigner acquires a state's nationality by marrying someone from that nation. In the USA for example,

a permanent resident who is married to a citizen is eligible to naturalization after just three years. Note, the higher the number of foreigners that get married to a particular nationality, the higher the proportion of that nationality in the population. On the other hand, in cases where dual citizenship is not allowed, one of the parties in that marriage loses their nationality. That way, the percentage of members with those nationalities declines.

2. **Legitimation or Adoption of Children:** When a child is adopted, they became part of the family of their parents. When the child is from another country or has a different nationality from their parents, they automatically gain the nationality of their parents. The adopted child can rightfully earn the nationality of that country if one of its adoptive parents is legal nationals. However, the child must be below eighteen years. Also, the child must have been admitted to that country legitimately. In addition, the parents must have followed the right channel to adopt the child. The adopted children may lose their former nationality in countries that do not support dual citizenship. The nationality that the child changes to, gains an extra member while their former nationality loses one member.

3. **Acquisition of Domicile:** One acquires domicile by showing intent to permanently or indefinitely reside in a country. Domicile can consist of domicile of origin whereby the child acquires domicile by reason of birth. In this case, a child born in a marital situation will assume the domicile of its father while a child born out of wedlock will assume the domicile of its mother. There is also the domicile of choice which is inferred by law. The law will give this if there is a change in the residence which is intended to be permanent. Domicile by operation of law is one that is imposed by the law for example when the parents of a child change their domicile, the child's domicile also automatically reflects that to parents and also when a woman gets married, she assumes the domicile of her husband although she also has the option of having an independent domicile. They may lose their former nationality decreasing its proportion in the population. Contrariwise, the nationality they gain earns a higher proportion of members (Boll, 2007).

4. **Appointment as a Government Official:** Once you are appointed as the government official of any country or state, you earn the nationality of that country or state. The official may lose their former nationality which results in a decline in the proportion of that nationality in the population.
5. **Resumption:** This is whereby one reinstates their nationality after losing it for one reason or another.
6. **By Subjugation:** This happens when a state is conquered and hence all its citizens acquire the nationality of the conquering country.
7. **Cession:** This happens when a state relinquishes and merges with another state and therefore the whole population acquires the nationality of the state which their territory has been merged.

2.8.3. Loss of Nationality

This is where one stops being a citizen of a nation either voluntarily or involuntarily. Involuntary loss of citizenship is whereby one's citizenship lapses automatically due to maybe one failing to take a certain action to retain his/her citizenship or withdrawal of citizenship by the country.

Voluntary loss of citizenship, on the other hand, is also known as relinquishment or renunciation is initiated by the citizen for example due to voluntarily serving in a foreign military or voluntarily naturalizing as a citizen of a foreign nation (Boll, 2007).

Different countries have different laws on nationality. Below are some of the most common ways in which nationality or citizenship can be lost in various countries:

- Being naturalized in a foreign country after reaching the age of eighteen;
- Taking oaths and formal declarations of allegiance after reaching the age of eighteen;
- Serving in foreign militaries that are hostile to your country or serving as commissioned or noncommissioned officers in foreign militaries;
- Taking up office, post or employment in foreign countries after reaching the age of eighteen and taking up the nationality of those countries;
- Taking up office, post or employment in foreign countries

after reaching the age of eighteen whereby an oath or declaration of allegiance is required; and

- Formally renouncing citizenship before a diplomatic or consular officer of a foreign state.

Remember, different nationalities may exceed the proportion of other nationalities in a single population.

2.9. POPULATION RELIGION

Religion is the belief and commitment to a certain being (Figure 2.10).

Figure 2.10. Different members of a population have different religions.

Source: https://www.msn.com/en-us/lifestyle/smart-living/what-these-14-religions-believe-about-god/ss-AAmvze2.

The proportion of followers of different religions within a population varies. Also, different populations are dominated by different religions (Holdsworth et al., 2013). The percentage of believers in a religion reduces and increases depending on various factors.

2.9.1. Factors That Influence Population Religion

1. **Economic Factors:** These have a massive influence on population religion. People migrate from place to place in search of greener pastures. The places they move to are inhabited by people that believe in one or various religions. People may even move to places whose populations believe

in religions different from theirs. Even if the main reason for moving to that place was to make money, these people have to interact with the other members of the population. Through their interactions, they get to learn about other religions. Consequently, they may change their religion. There are various reasons why people who migrated for economic reasons may want to change their religion:

To Earn Confidence among the Members of That Population: Once a person is bold enough to convert to the religion of another, they gain trust and confidence among the members of that religion. Additionally, they earn more clients to their business from the members of the population particularly from the members of the religion they convert to.

To Learn Business Strategies: This is a new locality, you may be required to learn the ways of life of the population in that area. Religion is one way of life that you may need to learn. Remember, some religious beliefs may inhibit the growth of certain businesses within the areas inhabited by their believers. It is thus essential to learn the religion of the population in that area. The best to learn a religion may be practicing it. The more the people that practice religion for business reasons, the higher the percentage of followers of that religion within the population (Holdsworth et al., 2013).

2. **Social Factors:** Social aspects also have an influence on the religion population. The social factors that influence religion population include:

- **Intermarriages:** When a person gets married to a member of a different religion from their own, they may be forced to convert. The religion they convert to automatically gain one more member while the religion they convert to lose one member. Remember, the more the converts a religion gain, the higher its proportion goes.
- **Interest in a Religion:** After associating with the members of a particular religion, a person may like the practices of the religion. They may then decide to convert to the religion. After they convert, that religion population increases in proportion while the religions they convert from experience a decline in proportion.
- **To Fit In:** In a locality where the inhabitants observe a particular religion, a person may feel the need to fit in society. One way of fitting in is by observing the same

religion as the rest of the members of that population. Once they convert to another religion that population gains more members that belong to that religion. On the other hand, the religion they convert from experience a decline in proportion.

- **Education:** As members of the population go through school, they learn about various religions. They also interact with members of different religions. The teachings they receive may draw them towards a particular religion and away from another. Consequently, they may convert into that religion adding one more member to that religion. The religion they convert from then losses one member.
- **Popularity:** A proportion of the population also becomes members of a particular religion with the aim of gaining popularity among family and friends. Some religions are considered to be classier than others. Such religion, therefore, tends to attract more members. The more followers they attract, the higher their proportion gets. Contrariwise, there are religions that are considered to be inferior. Such religions are unlikely to draw followers. Subsequently, such religions record low proportions.
- **Norms and Obligations:** When a child is born and brought up in a family that observes a certain religion, they automatically become members of that religion. The children may not have the liberty to choose the religion to follow until they are old enough. They, therefore, follow the religion of their parents as a norm. Others are obliged by their parents to follow the practices of that religion. The more the children born in a particular religion the more the proportion that religion has in the population. On the other hand, a religion that records fewer births, experience a lower proportion in the population.
- **Religious Beliefs:** Some religious beliefs also influence the religion population. Some beliefs influence the birth rates within the members of a particular religion. For instance, some religions discourage their believers against the use of birth controls. In such a religion, the birth rates among members are high. The higher the birth rates the more the percentage of the followers of that religion. Children born into that religion get absorbed in the religion increasing the proportion of its members. On the other hand, religions that do not limit their believers from using birth

control may record a lower birth rate hence a lower proportion of members in the population.

3. **Technological Factors:** Technology advances from time to time. In the recent past, there have been massive technological improvements that have massive effects on people's usual undertakings. Religion is one area that has been affected by technology. Technology has made information and teachings about various religions available to the general public. The teachings may draw a proportion of the population towards certain religions.

Some religions also receive negative criticisms in technological channels such as the media. The media channels that give teachings about various religions include televisions, radios, newspapers, and the internet. People may convert from these religions as a result of the negative teachings they learn about their religions.

The religions that people convert to as a result of technological influence experience a higher proportion of the population. On the other hand, the religions they convert from experience a decline in proportion.

2.10. Population Sex Composition

Population sex composition is also referred to as the population sex ratio (Figure 2.11).

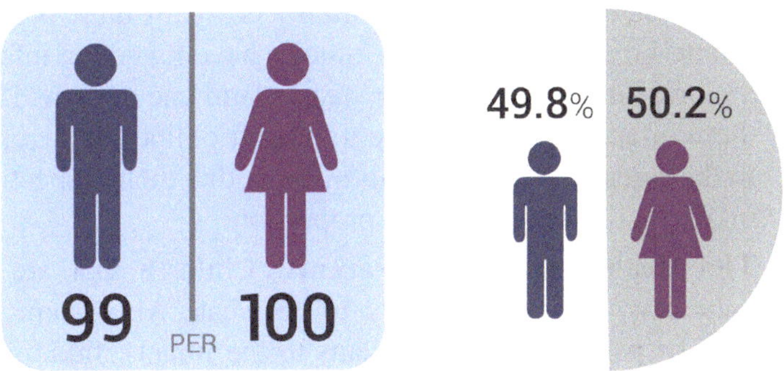

Figure 2.11. Sex ratio is the proportion of males and females found in a population.

Source: http://rssoarmm.psa.gov.ph/release/54742/factsheet/armm-population-sex-ratio-%28based-on-popcen-2015%29.

The sex ratio is reached by calculating the males in a population against the females. A population can have an equal proportion of both sexes. It can also have a larger proportion of one sex than the other. The sex composition of a population affects various fields in human life. Additionally, there are various factors that affect the sex composition of a population (Rosset, 2017).

2.10.1. Factors That Affect Population Sex Composition

These are the factors that lead to an increase of one sex more than the other. They determine which sex dominates the other in proportion within a population. Various scholars have studied the field of population sex composition. There are four major factors that affect population sex composition:

1. **Natural Factors:** These consist of aspects that happen without human influence. Such aspects influence birth and mortality rates. The aspects include:

- **Plural Birth:** It is when the number of infants born by one mother at a specific due period is more than one. Plural births may change the sex ratio within a population. For instance, if one mother gives birth to three girls in a plural birth and no mother gives birth to three boys within the same time, the proportion of girls will be less than that of girls at that particular time.
- However, the rate of infant mortality is usually more prominent in infants born in a plural birth. Studies have found out infants from a plural birth tend to have a weaker immune system. Therefore, these infants are more prone to pediatric illnesses that may lead to their deaths. If some or all of them die, their respective sexes may experience a change in proportion.
- **The Health History of Parents:** Children that are born to unhealthy parents are likely to be weak. Also, some diseases may be passed from the parents to the infants. Infants may not be strong enough to withstand the effect of the diseases and may die after some time. If for instance three girls were born to unhealthy parents and ended up dying. The proportion of girls in that population may decline.
- **Parent's Psychological Condition:** There are instances when children are born to psychologically disturbed parents. Some parents may also suffer psychological issues after they give birth

to their infants. Such kind of psychological stress is called post-partum stress. Psychologically disturbed parents may not be in a position to effectively take care of their babies. They may suffer from issues such as starvation, sickness and other condition that may harm them. These infants are more prone to early death than infants born to psychologically stable parents. Therefore, if for instance many boys are born to psychologically disturbed parents and ended up dying, the male proportion in that population declines.

2. **Environmental Factors:** These are that affect sex composition in a population are in various forms:

- **Climatic Environment:** Climate is a major factor that affects birth and mortality rates in a population. There are some hazardous climatic conditions that cause death to members of a population. Such conditions include drought, floods, and winds among others. In most cases, when climatic hazards happen, adult males survive than their female counterparts. Many children also perish alongside their mothers. If for instance, the higher proportions of children that perish are girls, then the boys in the population may surpass girls. Also, the male proportion may become higher than the female proportion if many women perish from the hazard. On the other hand, the climatic environment may favor the survival of humans. Some weather conditions in a certain place may favor practices such as agriculture. The population within the area then has enough food to survive on.

The mortality rate in such a population becomes lower. The birth rate may also go high. The sex ratios in the population are sustained due to lower death rates. The babies born in such a population are the key determiners of the sex composition. If more male infants are born, then the female proportion in the population becomes lower than the male proportion. According to Catalano et al., report, if the annual temperature in a region increases by 1 °C, there will be an increase in the number of males by 0.001%. They also conclude that colder climates lead to more termination of male fetuses.

- **Gestation Environment:** It refers to all the conditions surrounding a pregnant woman. Negative gestation environment such as psychological stress or scarcity of food makes the mother weak. The fetus within her may die. Studies have found out that most fetuses are less likely to survive when the gestation environment is unconducive. Therefore, if there is a drought in an area that has led to famine, expectant mothers may lack sufficient food. Consequently, the fetuses are at risk of getting terminated. In such a population, there are few fetuses carried to term. If for instance, most fetuses carried to term are females. Then, female sex composition in the population is higher.

3. **Social Factors:** These also affect population sex composition. Such factors include:

- **Sex Preferences:** These refer to embracing one sex while dismissing the other sex as inferior. In some communities, one sex is favored than the other. The preferences vary from community to community. Parents are pressured to give birth to babies of the preferred sex. In some cases, when the children born carry the other sex, parents end up giving birth to many children with the hope of giving birth to the preferred sex. In such communities, babies of the other sex may receive inferior care than their preferred counterparts. When babies do not receive the care they need, they may not survive. They may die at a young age. The preferred sex, therefore, stands a better chance of survival. The sex composition of the preferred sex is therefore likely to be higher than the other.

In the Philippine context, why are daughters valued in Filipino families, so much so that they are sometimes preferred over sons? Here is a sample of the most cited reasons for wanting daughters: "so that they can help in housework and family chores," (Bulatao, 1975, 1978), "as a source of assistance in carrying out household tasks" (Licuanan, 1979), "para makatulong sa bahay" (so she can help in the home) (Castiglioni, 1982) and take care of their younger siblings.. Only a tiny minority offer other reasons for wanting a girl-child, such as particular qualities of personality or behavior that she may have (Bulatao, 1975), or the fun, happiness, and companionship she can provide – especially to her mother (Bulatao, 1978). Macalandong, Masangkay, Consolacion, and Guthrie (1978) report that in the Maranao culture, female children are desirable because "they command a good dowry from a husband's family", although male children are still preferred because they can be their fathers' "army".

Families are known to generally ascribe incentive value to children-viewing sons as well as daughters in terms of economic utility and psychological security (Bulatao, 1975, 1978; Palejo, 1992)-but the traditional and stereotyped aspects of the woman's role are still heavily emphasized in the reasons given for a daughter's welcomed presence in the family.

3

POPULATION AGING

CONTENTS

3.1. INTRODUCTION

This chapter looks to review matters suitable to population aging. The literature is divided into four segments. The leading segment evaluates demographic transition as one of the reasons behind population aging. In the following section, population projections of mortality, fertility, and total population are analyzed. The third portion assesses indices, a couple of examples and imputations of population aging, as well as various responses by governments to population aging. The fourth and final portion presents an overview of the reviewed materials.

3.2. DEMOGRAPHIC TRANSITION'S RELEVANCE TO POPULATION AGING

In the simplest and most comprehensible terms possible, population aging can be described as the procedure through which aged individuals form a lion's quotient of a country's overall population. It can be expounded as a general term for the conveyance in a population's age composition to older ages as a result of demographic transition. Population aging can also be termed as demographic aging and warrants the surge in the older ages' percentage composition of the overall population. Older ages are assumed as 65 years of age and above. Among the key facets of population aging are (Figure 3.1):

- A population's median or mean age increase;
- An increase in the elderly population composition;
- A decrease in the population whose age group is less than 15 years of age.

Figure 3.1. Population aging.

Source: https://www.npr.org/sections/health-shots/2013/10/08/230175345/delaying-aging-may-have-a-bigger-payoff-than-fighting-diseases.

3.2.1. What Exactly Is Demographic Transition?

Demographic transition is simply a process involving the methodical shift of demographic changeable from an initial state to a distinct one. It can also be termed as the progressive adaptation from high rates of births and deaths to significantly low death and birth rates accompanying both economic and social changes facilitated by industrial modernization. Demographic transition has been established as the reason behind population aging in both developed and developing countries, with a gradual and rapid transition, respectively. Throughout the demographic transition, the decline in fertilities happens to be the key contributor to population aging as it proliferates the older ages population percentage. Also, the demographic aging process is further accelerated by mortality dwindling in old ages which significantly spirals the population of the aged.

In Africa for example, aging of the population has been effectuated by a rapid decrease in fertility rates as a result of various socio-economic progressions and developments including the surging adoption and employment of contraceptives. Additionally, the mortality rates of infants and children have sharply shot down thanks to the consistent improvements witnessed in the health care systems. This is especially true in medial income nations such as Tunisia, Algeria, Mauritius, Egypt, and Morocco. Nonetheless, according to Nabalamba and Chikoko (2011), relatively high HIV/AIDS prevalence in other African countries including Botswana, Zimbabwe, South Africa, and Swaziland has resulted in the deflation of adult age groups, leading to a relative increase in the aged/elderly population.

In a country like Kenya for instance, the primitive rates of births and deaths have steadily decreased over the last couple of years. Still, the rudimentary rates of death have been dwindling rapidly as compared to the relatively crude rates of birth, establishing that the nation is in the subsequent phase of demographic transition. Kenya's expedition of demographic transition was recommended by NCPD, 2013, through fertility decline by increasing contraceptive awareness and use, as well as reducing child and infant mortality through the enhancement of female enrollment in schools.

3.2.2. Projection of the Population

Population projection is simply an estimation of the future population involving the cogitation of various population changes facilitated by fertility and mortality, as well as migration. The demographic tendencies procured from the projections of the population play a very vital role in highlighting some of the immediate and subsequent policy challenges for the governmental directorate. Population projections are also vital for resource distribution and allocation, policy change, and advocacy as well as research, monitoring, and appraisal to establish whether or not the country in question is still on track in realizing national and also international development goals. As discussed below, there are various methods and techniques employed in the fertility and mortality projections.

3.2.3. Fertility Projections

Projected fertility levels have very crucial and significant innuendos for the coming populations' age structure, which includes the momentum of population aging. The United Nations generates deterministic projections of the total fertility rate (TFR). It then fragments the TFR that has been projected into age-specific fertility rates (ASFRs) using fertility time tables (Figure 3.2).

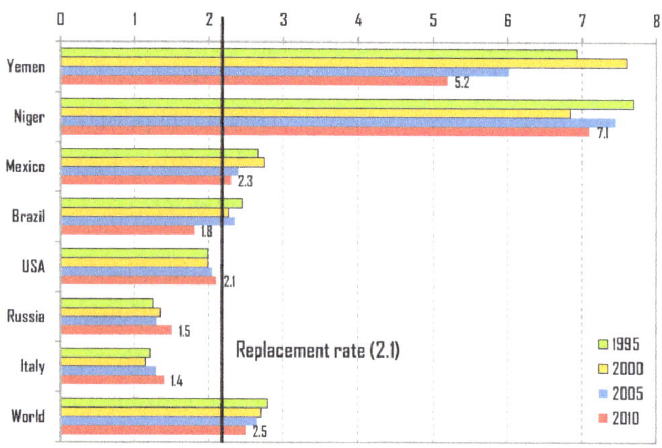

Figure 3.2. Total Fertility Rate, Selected Countries, 1995-2010.

Source: World Population Prospects, the 2010 Revision.

These are later merged with international migration and mortality projections through a projection technique referred to as the cohort-component. Deterministic projections are usually founded on often-biased expert opinion, on the foreseeable imminent itinerary of fertility, mortality, as well as immigration. Often, experts employing deterministic projections have been known to overstate ensuing mortality rates, resulting in a consequent trivialization of the aged population. Also, they predict or rather project low fertility if the recent past rates of fertility are low and relatively project high fertility in case the recent past rates of fertility are high.

Miller, by coming up with the random country model (RCM), made significant advancement on the deterministic projection. This RCM model employs the mutual experience shared by member countries of the UN as the foundation in the projection of following demographic tendencies as well as measuring their uncertainty. This model is exceptionally effective as it lays emphasis on the reciprocal experience of member countries and deliberately tries to limit expert knowledge on the country in question.

Alkema et al. (2010) have seen further advancements of the projection technique by employing the Bayesian projection replica in the production of TFR projections that are country-specific to all nations involved. TFR's evolution was disintegrated into three prime stages including; High fertility pre-transition, the fertility transition as well as low fertility post-transition. This method projects the subsequent TFR grounded on both the country's individual TFR history and also from the pattern of all other nations. Fosdick and Raftery further extended the Bayesian model to incorporate probabilistic projection of the composite TFR for any set of principalities through the molding of the correspondence between state forecast fallacy as a linear function of the invariable time covariates.

The recent approach of fertility projections has laid more emphasis on ASFR rather than TFR. Pandey and Singh analyzed the past and futuristic ASFR tendencies of India's state of Uttar Pradesh then employed the Gompertz prototype in the Bayesian framework in the projection of each age group's ASFRs independently.

3.2.4. Mortality Projections

The current highest standard of mortality tendencies projection and fitting is the Lee-Carter model. This model presumes that the dynamics of mortality tendencies over a while is discerned by a single constant, the mortality index. Understand that this parameter is extrapolated via the selection of an apt time series model.

Essentially, this Lee-Carter prototype delineates the age-specific death rate that is logarithmically transformed ($m_{x,t}$) as an aggregate of an age-set constituent that doesn't depend on time (a_x), and the product of a parameter that varies with time (k_t), the index of mortality, that gives a summary of the general mortality level as well as an auxiliary age-specific constituent (b_x)

that depicts the changing rate of the mortality index. Below is a mathematical presentation of the same.

$$\ln(m_{x,t}) = a_x + b_x k_t + \varepsilon_{x,t}$$

The final expression, $\varepsilon_{x,t}$, is the expression of error, that are intended to reflect age-set influences that aren't captured by kt.

The central appeal of this model is its series component linearity that is long-term, its mortality index, k_t. Nonetheless, this model has also faced criticism from Girosi and King. They insists that this model tends to disregard all data points except the first and the last one, hence resulting in age profiles that grow less smooth over time regardless of the tendencies existing in the factual data, deviating from any provided baseline in the end.

Lee and Carter's approach was specifically developed for U.S. 1933–1987 mortality data. Their salient recognition was that the United States federal chart mortality data have a history of following a fairly linear age track in the history, of course, with a couple of exceptions. Distinctive age sets are, similarly, highly correspondent over time. This technique has now been adopted and is being employed to a general idea, as well as idea-specific mortality data sourced from various countries and periods, all way beyond the brackets of its intended design applications.

Using Canada's and well as the United States' mortality data, Li, and Chan conducted coherent outlier scrutiny of the Lee-Carter model's mortality index construct. They established that its linearity holds notwithstanding various outliers detected in the mortality levels given rise to by wars and pandemics. Nonetheless, they advocate for the employment of a more mainstream class of time series models that are rather non-linear for more meticulous evaluation of the mortality index, k_t in mortality analysis yet to come.

3.2.5. Stable Population Model

The formation of a stable population is characterized by constant age-specified fertilities and deaths over a prolonged duration of time without any migration gains or losses on the population. This stable population model was unfolded in 1922, by Alfred Lokta, who observed the special scenario in which the crucial age-specified rates of fertilities and mortalities remain unchanged year after another.

The impediment of the stable population model is that holding the crucial rates constant is quite unrealistic as the rates of birth and death fluctuate over time. Nonetheless, the theoretical correspondence between mortality, fertility, and the structure of age in stable populations subscribe to the comprehension of the structure and growth of historical populations. Additionally, it can be employed in the generation of demographic estimations when pragmatic data are deficient or of inferior quality.

UNPD has been employing this stable population technique in making its population projections. Similarly, many states and scholars have adopted and employed this model in generating their population projections.

3.2.6. Matrix Projection and Cohort Component Methods

The most commonly-employed population projection technique is the Cohort component method. This approach is founded on a stable population replica. However, the key difference is that this approach does not necessarily presume constant survival and fertility rates. Additionally, this method does not assume fixed net migration numbers.

The downside to employing the cohort component technique is that it highly depends on sound birth, death as well as migration statistics and surmises that the rates of survival and births and the net migration estimates remain unaltered throughout the entire projection period.

Quite several assumptions on the rate of growth are also made, including low, medium, and high anticipation rates for any coming changes, resulting in relatively broad disparities. Another notable point is that the projections' integrity has also been questioned; this has been true especially for the projections unfolded by the UN. In 2001, Keilman discovered that the UN projections are more accurate for shorter than longer projection periods; more reliable for larger than relatively smaller states or regions. He also unveiled that the young and elderly projections tend to be rather unreliable as compared to predictions of the intermediate age sets and the projections' accuracy vary considerably from one region to another.

The pattern and path specification of age-specified rates is also quite intricate, emanating projections that are hardly replicated without accessing proprietary computer program or software used by the crew that conducted and came up with the projections.

Matrix projection, on the other hand, is a specific script for the Cohort component method expression. Numerous demographic occasions feature the employment of the matrix projection technique. Matrix projection methods were instigated in the 1940s by Lewis, Bernardelli, and more predominantly Leslie, who developed matrices application fully. Application of matrix models was primarily neglected until the mid-1960s when they were rediscovered by some human demographers and ecologists.

Projection matrix was simplified by Keyfitz. As from the second row frontwards, only non-zero components are contained sub-diagonally, which represent the probabilities of passing from the current age set to the next one. Compositions of fertility probabilities at distinct ages as well as survival probabilities for infants given birth to the projection duration. As time went by, matrix projection has been enhanced by the use of surviving probability instead of zero as the final element in the last, open-ended group of age. The downside of this matrix projection technique is its assumption of zero migration all through the duration of projection.

3.3. POPULATION AGING MEASUREMENT

Following Grigsby and Olshansky, there isn't a single demographic quantify that can be rendered as the perfect compute of population aging. There are numerous demographic indicators of population aging. They include the number of people aged 65 and above, the proportion of the overall population aged 65 and above, aging indices, median age, proportion of individuals aged 85 and above, life expectancy at 65 years of age, and old-age reliance/ dependence ratio (OADR) as well as potential support ratio.

In most states, population aging is quantified by a percentage increase of either the elderly folks of the retirement age (60 or 65 years depending on the countries policies) or persons reaching the age of 60. The percentages are pretty important due to their aptness on the issue of dependency, sporadically quantified as the elderly's ratio to other adults, or rather the pensioners' ratio to taxpayers. The tally of older people is also of great significance when it comes to the demographic survey of the aged. This is especially due to the arising concerns of how many among them will need support and help in terms of housing, pension, and also services on health and welfare.

Often, the median age is employed as a premise for a population's establishment as young, old, or aging. A population that features a median age of 19 years and below is recognized as 'young,' between 20 and 29 years as 'intermediate' while a population featuring a median age of 30 years and above is described as 'old.' This median age is quite important due to its instinctual appeal when it comes to overviewing a population's age structure hence used in the comparison of two or multiple populations. It also facilitates the quantification of population aging without necessarily having to establish what age is considered old.

According to Shryock and Siegel, the best quantification of population aging is the aging index. Aging index, described as the ratio of the aged to the youth, is considered the best measure of the same because of its inclusion of two sub-groups that tend to mostly change in the period of demographic transition. The age sub-groups being referred to are of the individuals aged below 15 years and of those above 64 years of age.

Cuuaresma establishes that the most frequently applied aging index is OADR. This is because it features very important imputations for the social security systems' solvency which include public health and pensions and also for the request of individual reassignment to older members of the family from the working population age. The ratio defines and establishes the load of the aged non-working population place on the working population.

Nonetheless, it is argued by Sanderson and Scherbov that just equivalent chronological age is compared by OADR across periods, perpetuating the conjecture that an individual currently aged 65 has similar characteristics as one who aged the same 65 years, maybe 60 years ago. They propound that the very vital roles played by the upsurge of life expectancy in the universal aging process are ignored by OADR and advocate for the employment of prospective old-age dependency ratio (POADR), which considers the abiding lifetime (probable age) rather than the chronological age. However, Bloom et al. have criticized POADR arguing that despite the life expectancy increase, people essentially don't work to overdue ages and if it so happens, they relatively become less productive.

In 1989, Grigsby, and Olshansky observed the dependency ratio as a very significant index for policy purposes but not quite as a population aging indicator. They contend that this dependency ratio is solely influenced by population aging and that because age isn't an ideal index of economic activity, this dependency ratio perpetuates the cynical stereotype that folks aged above 65 years seem to depend on the ones between the age brackets of 15 to 64 years.

3.4. POPULATIONS OF THE AGING

As demonstrated by the population of the elderly (aged 60 and above), the population of the world has been steadily aging. The estimates provided by UNPD indicate that the elderly population percentage was at 8% in 1950 and rose to 11% in the year 2010. The latest estimates provided by the same organization, has the same elderly population at 12.3% in 2015. It has also been projected that the same population will be at 16.5% come 2030, and a surge to 21.5% in 2050.

Throughout the world, the elderly population has been observed to increase with Africa as the exception, where it constantly held at 5% from 1950 to 2010. Nonetheless, this population increased to 5.4% in 2015 and as per the projections, it is expected to rise to 6.3% in 2030 and up to 8.9% in 2050. In Asia, the same population was at 7% in 1950 and 10% in 2010. This elderly population grew to 11.6% in 2015 and is projected to grow to 17.2% in 2030 and eventually to 24.65 in 2050. In the case of Europe's elderly population, it was at 12% in 1950 and 22% in 2010. This population elevated to 23.9% in 2015 and is projected to proliferate 29.6% in 2030 and eventually to 34.2% in 2050.

As identified by UNPD, Japan is the country that features a lion's share of the elderly population which was at 33% in 2015. Germany came in next at 28%, Italy at 28% closely followed by Finland at 27%. UNPD projections indicate that Japan's elderly population will be at 37.3% in 2030 and eventually surge to 42.5% in 2050.

The world's most populous nations, India, and China, whose elderly population is also projected to increase, are not left behind. China's elderly population was at 15.25 in 2015 and projected to increase to 25.3% in 2030 and 36.55 in 2050. Due to these countries' populous nature, translating their respective percentages to numbers is bound to yield sizable elderly populations.

In Africa, the country with the highest elderly population composition as of 2015 was Mauritius. Its elderly population was marked at 14.7% and is projected to hit 23.3% in 2030 and eventually 30.6% in 2050. The runner's up in this section was 11.7% in 2015 and projected to reach 17.7% in 2030 and eventually 26.5% in 2050. In Kenya, the population in question was at 6% in 1950, declined to 4% in 2010, increased once more to 4.5% in 2015 the projection estimates have it at 5.5% in 2030 and eventually at 9.6% in 2050.

Over the last couple of years, life expectancy has steadily increased. Keep in mind that this is also an indicator of population aging. UNPD established the world's life expectancy in 1995 at birth, for both males and females were 64.5 years, while, during the same period, Kenya's life expectancy was at 57.3 years. By the year 2010, the world's life expectancy had grown to 68.8 years of age, while Kenya's had slightly decreased to 56.5 years. In 2015, the world life expectancy at birth was 70.5 years while Kenya's life expectancy at birth stood at 60.6 years. Projections have the world's life expectancy at 71.7 years in 2020, expected to rise to 73.7 years in 2030 and eventually to 77.1 years by 2050. For Kenya, this life expectancy at birth is projected to be at 63.3 years in 2020, expected to rise to 66.0 years in 2030 and eventually shoot to 71.7 years in 2050 (Figure 3.3).

Figure 3.3. Aging population.

Source: https://www.healthline.com/health-news/how-should-countries-care-for-aging-populations.

3.4.1. Population Aging Implications

In 1980, Clark, and Spengler outline that the problem faced by population aging is supplying ample elderly economic security. On the contrary, it is argued that the potential of growth is vanquished by population aging through decreased domestic savings, diminished labor inputs and also pressurizes households and public finances following a surge in medical expenses as well as augmentation of pension burden.

The American National Intelligence Council argued that states facing an evolution in the aged population face probable dwindle in economic productivity and relatively slow aggregate GDP growth or worse yet, stagnation. Cipriani also outlines that population aging, following a decline in fertility and longevity, compromises the pay-as-you-go practice's solvency by creating a significant increase in the capital-labor ratio which results in the increase in output per capita and also pension pay-outs. Here are some of the notable impacts of population aging:

- An upsurge in the dependency ratio;
- Increased government expenses on pensions and healthcare;
- Heightened taxes for the working class;
- Deficiency of workers;
- The accrued market for services and goods linked to the elderly; and
- Diminished capital investment due to augmented pension's savings.

The amplified concerns by policymakers, especially in the U.K. and U.S. over implications of population aging have been blown way out of proportions by the philosophical aversion of public expenditure on health care and pensions which may also aggravate intergenerational squabble if used as a premise for the social policy development. Nonetheless, Bloom et al., gives some hopes that demographic aging features various challenges that are quite formidable, though not insurmountable. They point out that some behavioral retorts can attenuate these effects of age structure.

In Kenya for instance, extended families, in which the needs and requirements of the elderly are often met is disintegrating slowly, even though most of them live with other family members. Additionally, elderly folks suffer from a variety of health complications but are however not able to cater for their healthcare. This also applies to those receiving pension thanks to the meager amounts they are paid. Consequently, the older folks in Kenya are regarded as vulnerable by the NCPD and require a lot of attention and taking care of.

3.5. GOVERNMENT RESPONSES TO POPULATION AGING

With the protracted proliferation of the aged populations, various challenges are presented to policymakers in diverse communities. The constant increment of these challenges has resulted in the establishment of various implements geared at addressing the said challenges at territorial, state, and international levels.

In 1982, the first-ever international implement meant to guide the formulation and structure of policies on the elderly was established, the Vienna International Plan of Action on Ageing. Its adoption occurred during the first World Assembly on Aging conducted from July 26 to August 6, 1982 in Vienna, Austria. The objective of this instrument is to strengthen governments' as well as civil society's capacity to effectively handle population aging and be able to comfortably address dependency needs and developmental potential of the elderly. Additionally, this act looks to enhance international and even regional cooperation. Nonetheless, note that during that time, demographic aging was principally a concern for countries that were developed.

The second aging implement is the 2002 Madrid International Plan of Action on Aging (MIPAA) endorsed in April 2002 during the World's 2nd assembly on Aging conducted in Madrid, Spain. MIPAA symbolizes the turning point on how key challenges on the development of a society that accommodates all ages comfortably are addressed by the world. Its sole objective is promoting and endorsing a developmental technique of approaching population aging through the mainstream of the elderly folks into the state as well as international development policies and plans throughout all sectors. It is aimed at three key sectors that include: older folks and development; enhancing well-being and health into later ages; advocating for enabling and supportive environs.

Soon after the endorsement of MIPAA, that July 2002, the African Union established the African Union Policy Framework and Plan of Action on Aging. The framework of the policy seeks that every single member state recognizes the basic rights of the elderly folks and have themselves committed to abolish any forms of age-based discriminations. It also looks to fortify that the elderly's rights are well protected by apt legislation. As a follow-up to this implement, the African Union's ordinary 26th assembly conducted in 2016 at Addis Ababa, Ethiopia, assumed the African Charter concordat on Human and People's Rights on the Elderly's Rights in the continent. Called for by the concord is the enactment of the doctrines of diverse International conventions, declarations, and implements dealing with aging issues. The African Union and United Nations (UN) aging instruments have since been domesticated by Kenya. In February 2009, the Kenyan parliament implemented a National Policy on Older Persons and Ageing aimed at providing a detailed framework for guiding matters of aging and older folks in development programs and processes as well as appraises other sectorial blueprints. In 2010, the older people's rights were embedded into the Kenyan 2010 Constitution. The constitution's Article 57(d) acknowledges older people as a special interest class for protection and advocates that they receive decent support from their household members and the government as well to enable and help them lead dignified and respectable lives. In the 1st month of 2014, the policy on the elderly was audited and aligned with accordance to the 2010 Kenyan constitution via a consultative procedure that conformed to the requirements of the constitution. A comprehensive and detailed framework is provided by the policy that looks into some of the unique challenges faced by older persons in Kenya. It also recognizes the elderly as palpable participants and right holders according to Article 57 of the said Constitution. Among the directions taken by Kenya in terms of the said policy improvement in handling population, aging includes the 2009 increment of civil servants' retirement age from 55 to 60 years of age due to the budgetary curtailment in the payment of their pension. Similarly, the civil service pension plans in place that are non-contributory are being transformed into contributory retirement plans (Figure 3.4).

Figure 3.4. In July 2012, a welfare scheme initiated by the Kenyan government pays a Ksh.2000 monthly stipend to the poor elderly people all over the country. This scheme has since been improved and enacted under the National Safety Net Program (NSNP) to make the monthly payouts as non-contributory pension to all registered folks aged above 69 years.

Source: https://www.kenyans.co.ke/news/42982-inua-jamii-how-social-protection-cash-programme-works-kenya.

3.6. A SUMMARY OF THE CHAPTER

Population aging, also referred to as demographic aging is a general term used to describe the shifts in a population's age structure towards later or rather older ages. It is a result of demographic transition that involves the decline of fertility, as well as mortality rates in response to economic and social changes as well. Ultimately, the population's life expectation augments, resulting in significant high proportions and numbers of the elderly in the entire population.

The most commonly employed population aging indicator/measure/quantifier is the percentage or rather proportion of the overall population aged 60 or 65 years and above depending on the policies in place. Other quantifiers include numbers of the elderly, aging index, and the median age. Potential support ratio and dependency ratio, which are often influenced by demographic aging, are also realized to assist in planning. Demographic aging quantification in most states is based on projections conducted by the UN. Nonetheless, it appears that the UN projections of the young and old are far less definitive than their projections of intermediate age sets. The UN employs deterministic techniques in TFRs projection which is then disintegrated into ASFRs by the employment of fertility timetables. Finally, they are integrated with the age-specific mortality projections as well as migration projections via the employment of the cohort component projection technique resulting in projections of the medium variant populations.

Deterministic projections are usually founded on adept opinion, which is, however, often biased regarding the probable future course of fertility, mortality, and even immigration. Probabilistic projections, specifically the Bayesian models, have been developed to help curb these expert biases. Employments of nonlinear time series structures and structuring of age-specified mortality and fertility rates have been suggested to help overcome these biases.

The model most commonly employed in the generations of population projections is the stable population model, the cohort component technique to be more precise. However, this method tends to be quite complicated and complex as it holds multiple variables constant. The preferred method is the matrix projection which allows the replication of the matrix with variations for the critical rates.

Population aging has multiple implications on all features of the human life that includes social, economic, political, as well as cultural. It tends to suppress the potential to grow due to various relative challenges. Several programs, policies, and protocols on the elderly at regional, state, as well as international levels have been established to help deal with these arising challenges. Principally, governments have responded by increasing taxes and retirement ages and also increased investment in social welfare and pension schemes for the elderly.

4

HUMAN MIGRATION

CONTENTS

4.1. INTRODUCTION

Migration is the movement of people from one location to another with the goals of staying, forever or interim at a fresh place. The movement is frequently over long intervals and from one location to another, but domestic migration is also feasible; actually, this is the superior form worldwide. People may move as individuals, in household units or groups of large people.

A person who migrates from their residence due to natural catastrophes or civil rivalry may be construed as an alien or refugee or, chiefly within the same nation, as an internally displaced people. A person looking for refuge from religious, political or other ways of discrimination is usually construed as a shelter seeker. The difference betwixt spontaneous (running political rivalry or natural calamities) and non-spontaneous migration (commercial or migration of labor) are hard to make and partly idiosyncratic, as the instigators for migration are often comparable. The World Bank approximated that, as of the year 2010, around 16.3 million migrants certified as refugees. This number rose to 19.5 million by the year 2014. At levels of approximately 3%, the percentage of migrants among the population in the world has continued unusually constant over the past five decenniums. Movements among the nomads are mostly not viewed as migrations as the migrations are always intermittent; there is no objective to reside in the new locality. Interim movements for the reasons of tourism, travel, pilgrimages, or the drive is also not viewed as migration, in the absence of an objective to reside and home in the visited localities.

4.2. WHY PEOPLE MIGRATE?

Peoples have many objectives about why they may want to migrate from one locality to another. These speculations may be commercial, communal, environmental, or political. For movement to take place there are usually drive aspects and drawing aspects at work.

Driving aspects are the speculations that make someone choose to migrate. This is their knowledge of life in one place which gives them better speculations to migrate from it. Mostly driving aspects are adverse things for instance crop failure, unemployment, flooding, droughts, war, poor opportunities education or poor amenities and services.

Drawing aspects, on the other hand, are the anticipations that draw people to the fresh localities. They are usually conclusive things for instance as job opportunities, better living standards, better healthcare or better education (Ness, 2014).

The essential aspects which captivate people to migrate can be categorized into five classes. They are:

- Commercial aspects;
- Population aspects;
- Communal aspects;
- Civil aspects; and
- Other aspects.

4.2.1. Commercial Aspects

The majority of the investigation shows that migration is captivated by economic aspects. In advancing nations, low income from agricultural, unemployment from agricultural and also underemployment are seen elementary aspects pushing the migrants towards advanced localities with greater opportunities in job. Therefore, nearly all investigations coincide that most of the migrants have migrated in search of better opportunities commercially. The elementary commercial aspects which inspire movement may be farther categorized as 'Push Aspects' and 'Pull Aspects' (Ness, 2014) (Figure 4.1).

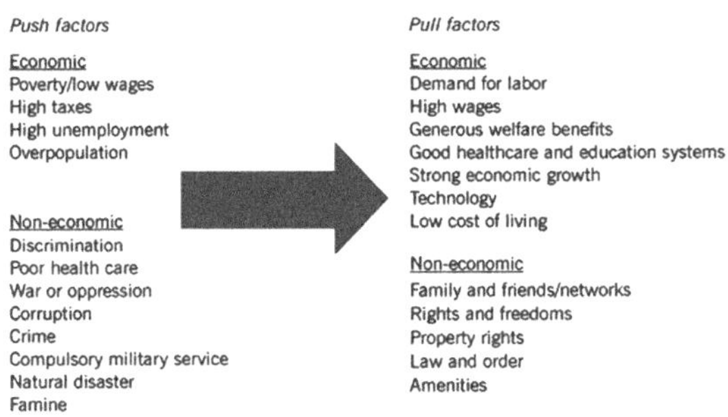

Figure 4.1. Push and pull aspects of migration.

Source: https://wol.iza.org/articles/demographic-and-economic-determinants-of-migration/long.

The push aspects are aspects that enforce a person, because of adverse objectives, to leave that locality and go to some other locality. The general push aspects are small productivity, unemployment, and under advancement, poor conditions of the economy, inadequate favorable circumstances for development, depletion of natural assets and natural disasters. Establishment of capital-exhaustive production methods into the agricultural sector and automation of certain systems decline the requirement of labor in rural localities. The non-capability of alternate beginnings of income in rural localities is also an essential aspect of migration.

The drawing aspects are aspects that favor the migrants to a locality. Favorable circumstances for good employment, wages that are higher, facilities, better conditions of work and alluring comforts are drawing aspects of a locality.

4.2.2. Population Aspect

The distinctness in the growth rates of the populace of the different localities of a nation is a stimulus in the domestic migration. Fertility and the general acceleration in the population are commonly higher in rural localities which drift the populace towards the city. Other essential demographic aspects in domestic migration are marriage due to females is used to pursue their companions.

4.2.3. Communal Aspects

Social and cultural aspects are also essential parts of the migration. Consistently conflicts in the family, the quest for freedom, also lead to migration specifically, of those in the younger generation. Enhanced facilities of communication, for instance, transportation, television impact, good communication network, the cinema, the urban familiarized education, and resultant transition in approaches and values also encourage migration.

4.2.4. Civil Aspects

Occasionally even political aspects promote or hinder migration from locality to another. After 1948, the majority of rural people moved to urban due to safety issues in Myanmar. Therefore, the political environment approaches, and individual opinions of the people affected the migration of people.

4.2.5. Other Aspects

Other aspects, for instance, the presence of one's household and friends in urban localities, want to accept education which is feasible only in urban localities are aspects of promoting migration. The proximity of contacts of culture, diversity in culture, great endurance, individual approaches is also related to migration.

1. **Circumstances in Agriculture:** Positions of labor surplus emerging from deficiency of cultured land, unfair land allocation, small productivity in agriculture, high density of population and the consolidation of the rural economy nearly restricted on agriculture commonly lead to acceleration in outmigration. This mixture of aspects makes a "drive" that is confronted more often in weak surroundings.
2. **Drought:** It is the standard "drive" affecting millions of people more so in Africa sub-Saharan and Southern Asia (Figure 4.2).

Figure 4.2. The migration of people in a delicate and unfriendly environment can be seen as one of the major idiosyncrasies and history drivers.

Source: https://www.ecology.com/2014/12/23/migration-merits-place-main-stream-climate-debate/.

Scarcity-prone Sudanese and Sahelian rural economies have tenacious attitudes of trade and migration of labor. It is approximated that the Nouakchott population in Mauritania increased forty times betwixt 1965 and the end of the 1980s due to prolonged drought and livelihoods loss. The migration will probably become even more essential: a European network of food security investigation of employment and mobility of labor in Ethiopia indicates that migratory labor is probable to play an accelerating essential part as a coping way for food-anxious rural homes (Ness, 2014).

3. **Water-Logging:** It can also be a cause for movement. In areas of Pakistan, the unrestrained use of irrigation water has culminated the waterlogging and salination of lands. This has driven to the dropping of crop outputs, which in turn has led to migration by destitute households.
4. **The Pressure of Population and Fragmentation of Land:** Having insufficient entry to land in a most importantly agrarian community leaves the landless with numbered substitutes to migration. In some Latin American nation's entry to land is so narrow that almost all destitute young people see migration as their major and possibly only livelihood alternative. The PPA in Ecuador effectively states that "The cries of destitute people in Ecuador are cries of people on the migration, traveling to new localities in hope of boosting their livelihoods."
5. **Erosion of River-Bank:** In Bangladesh, few districts are overwhelmed by the erosion of river banks which has been an essential push of migration. Conferring to an investigation administered in the late 1980s the number of people touched yearly by the erosion of riverbank in the areas of delta was 1 million. The investigation found that approximately 11% of the settlers in Dhaka came from the Barisal district and another 31% from Faridpur (both districts are super prone to the erosion of riverbank) (Figure 4.3).

Figure 4.3. People become homeless after riverbanks become eroded.

Source: https://www.observerbd.com/2014/10/16/48862.php.

6. **The Poverty of Mountain and Economies of Forest:** In the destitute mountainous locales of Southern Asia low productivity in agriculture, poor approach to credit or other pre-necessities for variety and high densities of the population make the situations for outmigration. A fresh rise in migration has been announced from Uttaranchal by Mamgain as the delicate mountain environs can't be sufficient for the rising populace. The destitute mountainous districts of Nepal also have bigger rates of outmigration. More or less the same aspects make a drive from most forested localities where population pressure has risen and CPR grounded occupation has become short-lived (Mangalam, 2015).
7. **Other Drive Aspects:** Separate from environmental drive aspects is the reduction of public sector jobs and complete inactivity in solemn sector job making. In Middle East and North African (MENA) nations, migration has been brought by a fast advancement of their forces of labor, high unemployment rates heavily rely on the public sector for job making (which is now being reduced under structural arrangements). Approximate 70% of the destitute in the MENA locale live in rural localities. Because of the 1970s, the employment share in agriculture has decreased faster but production and other industries have not risen proportionately making a current of rural-urban migration in all MENA nations, leading to fast urbanization and, in turn, to the conversion of urban areas into crowds of jobless people.
8. **Urbanization:** It has been a considerable driver of domestic migration in the majority of the nations and has passed other aspects in many Asian locales.

Urbanization rates affect rural-urban wage disparities: a rise in the labor demand in urban locations can drive urban wages up and raise migration.

Rural-urban disparities in moderate incomes rose in many East and South Asian nations during the 1990s, chiefly in China and fell in the majority of African nations. Antithetical to common wisdom on urbanization and migration, high migration rates (long-lasting and interim) into urbanized locations have progressed irrespective upcoming levels of (formal) joblessness and continual urban destitution. The clarification lies in the progressing urban casual sector which represents a meaningful draw (Mangalam, 2015).

In inclusion, urban localities give many commercial favorable circumstances to rural people for changing jobs fast and becoming rising movable with a very small asset-base and artistry. Even if wages in urban are not bigger, a job is available more commonly than in sustenance agriculture.

9. **The Fresh Moment in Agriculture:** Because the time that high output collections, guaranteed irrigation, and agrochemicals were brought in through the advancing globe in the 1960s, high agricultural productivity pockets have become a magnet for destitute migrant workers from destitute localities. This advances to be an essential draw because of the wage marked disparities in many nations even today: for example, there is much rural to rural movement betwixt upland and lowland Vietnam. Workers go from the dully populated delta of Red River in the north, to the highland of central boundary which had until very currently a rising export familiarized agro-economy majorly in coffee. Likewise in Ethiopia rural-rural migration has risen since 1991 where young men from densely populated localities go to irrigated lands for 3–4 months.

10. **The Calm of Constraints on Population Migration:** The loosening of constraints on population migrations has been an essential objective for the rise in flexibility in some nations. The best popular example of this is China. Pre-reform migration was restrained by the nation over an intricate system of employment and residentiary constraints, leading between which was the Hukou arrangement. This has been increasingly broken since the late 1970s and has, together with other aspects named earlier, resulted in rural-urban composition with other aspects named previously, pave way to increased rural-urban migration. Like increments in migrations have been noticed in South Africa pursuing the ban of constraining legislation at the end of discrimination. The other well-written example is Ethiopia where the movement of labor which had come to a stalemate under the Derg rule has started since 1991.

4.3. TYPES OF HUMAN MIGRATION

4.3.1. Internal Migration

Internal migration or domestic migration is the migration of humans in one geopolitical entity, usually; a country's domestic migration appears to be cross for academic and for economic advancement or due to natural catastrophes or civil commotion. Inter-boarder migration often happens for political or commercial objectives. A common tendency of migration from rural to urban localities, in a method characteristic as urbanization, has also created a form of domestic migration (Ness, 2014).

The population of Iran's internal migrants reached 684,000 last year, of whom around 567,000 resided in urban areas and 117,000 in rural areas (Figure 4.4).

Figure 4.4. Internal migration.

Source: https://medium.com/@Goodsvejk/internal-migration-and-group-conflict-422c1f214dfe.

1. **Incentives of Domestic Movement:** A bigger practical composition tries to examine which changeable clarifies most outstandingly the amount and order of migration internally. A few variables are better prognosticators in many nations.

- **Period of Animate Experience:** Migration is the greatest natural among immature employees. The ideal of human capital supports an elementary answer for this arrangement. Workers that are old have a less duration over which they can accumulate the rebounds to the migration financing. The less period conclusion period declines the net benefit to migration and therefore declines the chances of migration.

- **Academics:** Migration is more natural among highly knowledgeable employees. This interaction could begin due to more educated employees may be more effective at studying employment favorable circumstances in different markets of labor, thus decreasing migration expenses. It is also attainable that the geographic locality that makes up the appropriate markets of labor is bigger for highly knowledgeable workers. Examine, for instance, the market of labor encountered by professors of colleges. There are small industries in any likely city and skills of professors are greatly portable throughout universities and colleges. In effect, professors of the college market their artistry in a state and or even a global market of labor.
- **Interval:** Vast distances avert migration due to bigger distances indicate bigger migration expenses. The association betwixt interval and movement was an elemental part of the 'gravity ideals that ruled the composition before the establishment of the framework of the capital of human. The gravity deal assumed a precise collaboration betwixt movement and the capacity of the terminal and ancestry localities, as well as an indirect collaboration betwixt migration and interval.
- **More Changeable:** The confirmation connecting movements and other changeable is more enigmatic. Some investigations indicate that people who are not employed are more probable to migrate than people who aren't and that the flexibility of movement concerning the original locality's rate of unemployment tends to be stronger for unemployed workers. These investigations, however, originally ignore that both employment and migration weaknesses are internally concluded. Another analysis investigates if the differential wage betwixt the terminal and origin localities has a conflicting impulse on migration. This precise test of the model capital of human needs data on the wage current that the worker would face in both the origin and the terminal localities. As an outcome, the proof may be very keen on how the choice inclination challenge is controlled. Some investigations analyze

the relationship betwixt domestic migration and disparities in welfare socially advances administered by different jurisdictions. Although the proof is mixed, current proof indicates that 'magnets of welfare' attract specific typologies of people to acceptable regions.

4.3.2. External Migration

External migration means people migrating from one nation to another with dynamism in place of living long-lasting or for a long duration of time. In external migration, differentiate emigration which is the exit from one nation to another and immigration which enters the country of residents of another nation. In the populace of the nation moving external migration, not excluding immigration boosts the mechanical improvement of the populace (Figure 4.5.)

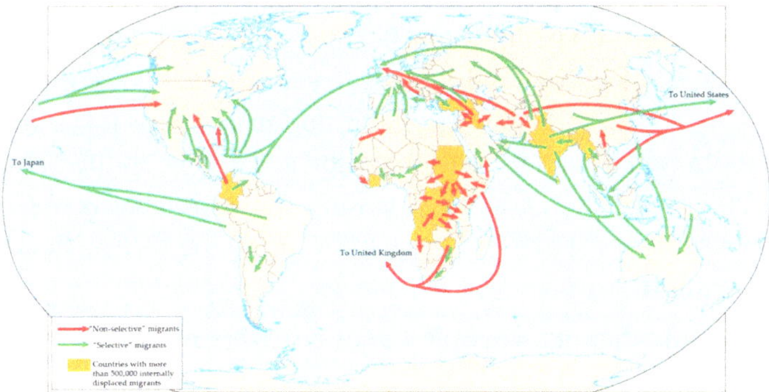

Figure 4.5. Selective and Non Selective Migrants.

External migration has continually happened. They played a big part in human advancement, specifically in modeling the structure of the population of many nations (Canada, South America, Australia United States, etc.). Reasons for external migrations may be political, economic, family, religious, household, natural, environmental, (related to natural dangers - droughts earthquakes, etc.). Economic aspects are the major and principal cause of migrants globally. For economic speculations, labor migration, i.e., job search and efflux minds (search-paid academic job by scientists and people of an imaginative line of work) happens when people cross nations' boundaries and live in the host nation for some minimum period. Migration occurs in many aspects. Abundant people leave their home nations to look for economic favorable circumstances in another nation. Others migrate to be with households who have moved or due to political circumstances in their nations. While there are several differing capability arrangements for classifying global migrants, one arrangement classes them into 9 groups: interim labor migrants; random, illegitimate, or unrecorded migrants; very skilled and job migrants; aliens; shelter seekers; compelled migration; relatives; rebound migrants; and lasting, low-skilled migrants. These migrants can also be categorized into two vast groups, long-lasting, and interim. Long-lasting migrants plan to build their lasting homes in a new nation and preferably obtain the citizenship of that nation. Interim migrants plan only to live for a definite duration of time; possibly until the end of a specific business of course or for the time of their job contract or a specific job season. Both typologies of migrants have a crucial effect on the economies and communities of the preferred destination nation and the nation of origin (Ness, 2014).

Essential obstacle related with external migration is the problem of aliens. Approximate 20 million people have become in the past few years compelled migrants because of global and inter-regional fights, civil combats, natural adversity, starvation, and scarcity. Each nation anyway faces obstacles of migration, coming up with migration policies. Relying on the demographics and requirements of the economy, national governments boost or confine the arrival of refugees. The following are some of the factors that encourage external migration:

1. Factors that push people out of a certain nation:
- Destitute Medical concerns leading to very high mortality rates;
- Inadequate employment opportunities;
- Primitive environment;
- Political phobia and instability;
- Fear of persecution and abuse;
- Religious maltreatment;
- Loss of prosperity;
- Natural epidemics such as earthquakes and floods; and
- Imperious and lesser chances of finding a companion.
2. Factors that pull people to a certain nation:
- High chances of getting an employment opportunity;
- Better living standards and enjoyment;
- Education;

- Better Medical responsibility;
- Security both to lives and property;
- Family associations;
- Lower infraction; and
- Better probability of finding a companion.

4.3.3. Emigration

Emigration is the act of leaving a native nation or locality with the reason to resolve abroad. Demographers investigate driving and drawing aspects for people to be driven out of 1 location and appealed to another place. There can be a crave to evade adverse situations such as inadequacy of land and employment, or biased analysis. People can be drawn to favorable situations accessible elsewhere. Running from adverse environments, being an alien and shelter seeker to get rank of refugee in an external nation, might point to long-lasting emigration.

Arrangements of emigration have been created by many communal, economic, and civil transitions across the globe in the past small 100 years. For example, millions of people ran away bankruptcy, assault, and civil unrest in Europe to resolve in the Oceania and Americas in the eighteenth, nineteenth, and twentieth centenary. Similarly, millions left Southern of China in the Chinese exodus in the nineteenth and previous twentieth centenary (Cummings, 2016).

Demographers categorize determinants at the ethnicity that drive out people, against those at the terminal that draw them in, as migration is concerned.

Few nations curb the capability of their nationals to move to other nations. For example, later in 1668, the Emperor Qing outlawed Chinese movement to Manchuria. The ruler organized the building of the Willow Palisade in 1681, an obstruction further which the Chinese were constrained from the impinging of Mongol and Manchu acreages. By the previous 1950s, the view of Soviet to administering global migration was also copied by North Korea, Mongolia, and China. Hardly North Korea continuously progressed to curbs movement, and sustains one of the draconian emigration censorships in the globe, even though some North Koreans still administer to banned China emigration. Some nations with fast emigration bans at a single time or other times involved, Egypt, Angola, Somalia, Mozambique, Iraq, Afghanistan, Laos, South Yemen, Burma, Northern, Cuba, and, Ethiopia (Figure 4.6).

Figure 4.6. A banner displaying a representation of the Cunard line's emigrant in 1913 liner RMS Aquitania sent off.

Source: https://www.art.com/products/p10096846-sa-i790682/rms-aquitania.htm.

4.3.4. Immigration

This is the worldwide migration of human beings into terminal nations of which they aren't nationals or where they don't have citizenship to reside or live there, exceptionally as long-lasting or accustomed nationals, or to take up the job as an interim or alien worker (Cummings, 2016).

1. **Understanding of Immigration:** London has become multicultural as an outcome of immigration. Black British in London in the year 2008 and British Asian offspring's exceeded white British children by approximate 1.5% in government-run schools.

One approach of immigration analyzes betwixt driving and draws aspects. Driving aspects indicate principally to the objective of immigration from the nation of ethnicity. In the case of commercial migration (mostly work migration), rates in wages are ordinary considerations. If the number of payments in the new nation exceeds the number of wages in one's origin nation, he or she may accept to move, as long as the prices are not heightened. Specifically, in the nineteenth centenary, the commercial progression of the United States accelerated immigrant inflow, and approximately 15% of the populace was alien delivered, thus constituting up an important aggregate of the effort of labor.

Non-commercial push aspects contain maltreatment (doctrinal and differently), constant crime, imperious, communal dangers to citizens cleansing, oppression genocide, in combat, and communal depreciation. Political reasons traditionally encourage alien circulation; for example, people might emigrate to run from a tyranny.

Obstruction to immigration arrives not only in the lawful pattern or governmental pattern; legitimate and communal obstructions to immigration can also be exactly capable. Immigrants when exiting their nation also consent everything common: their friends, household, culture, and backing system. They also require converting their goods, and they acquire the cost of leaving. When they appear in a new nation, this is rare with major ambiguity not to exclude looking for a job, where to reside, new laws, fresh norms of culture, issues of accent or language, discrimination possibilities, and other discharge behavior approaching them and their household.

The politics of immigration have developed accelerating related to other matters, for instance, security nationally and crime, exclusively in the western part of Europe, with the existence of Islamas a fresh prominent belief of divinity.

Investigations have indicated that some appropriate attraction association doorway for fewer immigration for their owned association and also immigration for other associations because they see causes of immigration, for instance, accelerated work challenge, as unfavorable when moving their association but advantageous when jolting other collaborations. An investigation in Europe in 2010 showed that "entrepreneurs are more expected to be for-immigration than workers, in that immigrants are speculated to challenge with workers who are previously in the nation. Differently, when immigrants are speculated to challenge with entrepreneurs fairly than workers, entrepreneurs are more feasible to be negative to immigration than workers. A study in 2011 studying the voting of United States representatives on migration policy suggests that representatives from more artistry work in many localities are more speculate to boost a free immigration rule towards the less skilled (Cummings, 2016).

Immigrants are driven to exit their nations of origin or descent for a collection of reasons:

- Ambition for commercial affluence;
- To look or involve in great pay jobs;
- To improve living standards;
- Reunion of the household;
- Running from animosity;
- Rivalry, conflict or common problems; and
- Clearly the appeal to transit one's life quality.

4.3.5. Return Migration

Return migration is the condition of the movement cycle, which was given the smallest concern, maybe due to most investigation assets are located in highly advanced nations, while most of returnees rebound to advancing nations. This is chiefly unlucky due to views about the system of rebounding to the nation of origin and beliefs towards rebounding migrants have an important force on migrants and their host societies. Thriftiness, often called remigration, is viewed by some writers as the last stage of the migration course, which farther entails the step development or choice to migrate and real migration stage or establishment of migrant the terminal nation has chosen. In the situation of transnationalism example in the analysis of migration, the rebounding is no longer seen as a final step of a narrow action because migrants prepare this across many visits and financial and communal payments. Reasons to migrate or to rebound forever to their home nation are not taken at the entity level but at the level of private groups or households, albeit the purposes for those members may be different. But experimental investigations depict that re-assimilation in the nation of origin is barely an easy one. Insensibility, migrants are not always knowledgeable of the fast transitions that can let them know their community of origin, even below acute communication with other household members or society of origin. On the other hand, the migrant rebound home after 10 or 20 years of stay in another nation is not the same as the person who left the nation at that time, his anticipations and those of the association to him can be diverse (Mangalam, 2015).

Some authors believe that migrants can be inspired to study overseas for boosting training traits and employability. Not all migrants rebound in the accommodating nations permanently, nor do they automatically cross over bigger durations of time. They can rebound, coming with the knowledge and entrepreneurial culture. Remaining migrants overseas have flourished in a notable bigger percentage than rebounded migrants to enter the market of labor before the first left. Approximately half (48%) of the people rebounded (contrasted with 38% of the staying migrants) have left without demanding the return or unable to enter the market of labor in the nation. The majority have found jobs overseas, but due to inadequate experience and competences, decrease in market of labor, have entered the most sensitive positions.

As a common rule, rebounding emigrants who have got financial accomplishments abroad are seen friendly and can even get a role model rank. Contrary, those who rebound home having declined in pursue of employment and stay overseas might be made to feel that they have not completed the anticipations of household and friends. Views are nuanced by their spontaneous or involuntary rebound.

The terminal migrants are engaged most often in the subordinate sector. In the absence of support from institutional, casual migration systems took over the accountability for the unification of migrants in the host nation, encouraging the subordinate sector employment, but giving narrow chances of occupational flexibility. The modest communal rank overseas opens favorable circumstances for communal ascent in the nation of origin, where they spend or invest the savings from migration and payments (Mangalam, 2015).

4.3.6. Seasonal Migration

Seasonal migration refers to the migration of people from one place to another during differing times of the year. The migration is essential due to seasonal decline in aspects such as the food availability, the temperature of the air, sunlight, and or even water become extreme in one locality, usually during the drought, making lineage and even continuation hard. An example is the migration of collection nomadic pastoralists from their summer homes in Africa where they move together with their cattle and families to areas which have pastures and when the drought is over in their home localities, they go back to their original localities.

For some communities, the seasonal movement cycle includes excursion that surpasses tens of thousands of miles or kilometers lengthwise. A route of migration can include taking rest and drinking stops, nevertheless, some communities exhaust their movement expressly. Attempts to defend migratory peoples include communal and private actions, chasing of shipping away from common routes of migration and agreements globally. The latter involves the meeting on migratory classes (Convention of Bonn) and the meeting of Ramsar on marshes. Seasonal migration feasible dates back at approximate 3,000 years, as documents of migrations written by the venerable authors have been recovered. Seasonal migration may happen over very long intervals. The Arctic course of tern betwixt the Antarctic and Arctic surpasses 22,000 kilometers, and the bar-tailed godwit moves unstoppably over 11,000 km from Alaska to New Zealand. Among whale classes, the blue whale moves from Arctic or Antarctic waters to its rearing areas in tropical waters in a movement that can take third a year, and in this duration, it does not intake anything. The course of the Pacific north-east gray whale betwixt the seaside waters outside Baja north of Okhotsk, and California the Sea betwixt Russia and Japan is approximately 10,000 km in dimension. Change of climate can affect intermittent migration. Birds movements have been affected for classes whose route takes them over deserts. Localities such as the Sahel location of Africa have evolved into drylands over the decades, declining the chances to break and refresh. In highly advanced localities of the globe, small pollution can also create challenges for people and birds that move at nocturne and use the heavenly bodies for direction (Cummings, 2016).

Intermittent migration, generally used correspondently with seasonal circular, short-term, and spontaneous migration, has been a discipline of much dialogue. Conferring to, all these migrations, usually temporary, repetitious, or recurrent, having the general objective of a temporary transition of stay, are cyclic in the description. Cyclic migrants pursue a circular direction and keep constant but interim absences from their localities of origin for more than a single day. Interim or Cyclic movement is a migration made for a short duration of time with the reason of rebounding to the place of usual permanence. An essential group of interim migrants includes intermittent migrants, who join the activity at various places conferred to seasonal requirements of labor (Mangalam, 2015).

4.4. TYPE OF THE HUMAN WHO MIGRATE

4.4.1. Immigrant

An immigrant is a person who comes to reside forever in a foreign nation. Finding that there are three typologies of immigrants: those who need the development of knowledge for their offspring; those who want to be rich and enjoy an affluent lifestyle; and those who want freedom. Those who want the development of knowledge for their offspring arrive at the new nation prepared to sacrifice everything to attain this objective. Once they report, they will get more than one job, commit their offspring to school (customarily private), and almost push themselves to early death in their endeavor to support this development of knowledge. They hold the majority of their original customs as they can and diffuse these into their offspring's as well. The difficulties that they come across are all in the name of love. These parents are to be applauded for their commitment but the offspring for whose well-being the parents are offering themselves, often find the circumstance very hard. The burden put on them to endeavor for completeness, the phobia and disgrace they feel due to they may let down their parents, and the accountability on their young shoulders--all these make them grow further their years (Ness, 2014).

On the other hand, the immigrants who want to be wealthy and have a moneyed lifestyle have another weight of their own. These people accept that cash is everything. They reason about urgent tangible achievement rather than durable ambition. They sometimes disregard their accountability as parents toward their offspring due to busy schedules demanding to make money. They mostly have more than one job and their family life has no balance. They accumulate to where work favorable circumstances are the biggest. Their offspring have no meaning in life; they never care about academics. All they care about is money-making.

Those immigrants who have come looking for freedom have a unique delight in reconstructing their lives. They have escaped from a nation with a restrictive system, a nation where human beings are not handled as people, but as commodities. They have run from a life that for some felt like existing in jail and for others was a prison. Regardless of the adversities, they have undergone in their nation of ethnicity, they still cherish it, even as they are advancing new pride and prefer for this new nation. But that nation will never absolutely be their home. They are continually looking out for a transition in their nation. They will grab the first moment to rebound to their nation.

They keep expecting for the announcement that their nation is transiting for the better, that there is no more socialism or that the autocracy has been replaced by an autonomous government; then they are ripe to return home. For a while, that nation was their shelter, their land of independence and it will continually be that, but it's not a condo. It is a place of shelter where one can be in impregnability (Ness, 2014) (Figure 4.7).

Figure 4.7. Immigrant in USA.

Source: https://fee.org/articles/understanding-why-immigrants-are-so-crucial-to-america-s-economic-growth/.

Most parents blend the connections of immigrants who want the development of knowledge for their offspring and those who want independence. They are running from socialism and have sought shelter in other nations. They are anticipating the announcement that will permit their return home. They are not nationals yet because the parents anticipate one day going back to their nation.

4.4.2. Refugee

A human being is a refugee inside the context of the Convention of 1951 as soon as he accomplishes the precedent included in the description. This would certainly happen before the time at which his refugee rank is formally resolved. Acknowledgment of his refugee rank does not make him a refugee but announces him to be one. He doesn't become a refugee due to acknowledgment but is accepted because he is a refugee.

The decision of refugee rank is a method that takes place in two stages. Firstly, it is important to assure the appropriate facts of the case. Secondly, the description in the Convention of 1951 and the Protocol of 1967 have to be enforced to the facts thus confirm (Mangalam, 2015) (Figure 4.8).

Figure 4.8. Refugee.

Source: https://foreignpolicy.com/2016/02/26/china-host-syrian-islam-refugee-crisis-migrant/.

The supplies of the Convention of 1951 describing who is a refugee consist of three parts, which have been termed mutually "inclusion," "cessation" and "exclusion" provision.

The inclusion provisions characterize the test that a person must attain to be a refugee. They form the considerable foundation upon which the decision of refugee rank is made. The purported cessation and exclusion provisions have adverse importance; the former shows the circumstances under which a refugee stops to be a refugee and the latter specify the circumstances in which a person is eliminated from the functionality of the Convention of 1951 although meeting the considerable test of the inclusion provisions.

The above specification is issued to supply a relationship with the past and to assure the progression of global security of refugees who became the responsibility of the global community at diverse earlier durations. As previously shown, these tools have by now lost much of their importance, and an analysis of them here would be of less real value. Nevertheless, a person who has been thought-out a refugee under the conditions of any of these tools is undoubtedly a refugee under the Convention of 1951. Therefore, a holder of a purported "Nansen Passport"4 or an "Eligibility Certificate" given by the Global Refugee Institution must be thought of as a refugee under the Convention of 1951 unless one of the cessation provisions has become appropriate to his case or he is expelled from the functionality of the Convention by one of the expulsion provisions. This also includes to an extant offspring of a lawful refugee.

Conferring to Article 1 A (2) of the Convention of 1951 the term "refugee" is used to any person who:

> "As an outcome of events happening before 1951 January 1 and owing to well-established phobia of being discriminated for mental analysis of race, nationality, religion, membership of a specific communal group or political attitude, is outside the nation of his citizenship and is not able or, owing to such phobia, is not willing to present himself of the security of that nation; or who, not having a citizenship and being outside the nation of his former established apartment as an outcome of such happenings, is not able or, owing to such phobia, is not willing to rebound to it."

4.4.3. Emigrant

Emigrant means all national of any country who contemplates emigrating, or emigrates, or has emigrated but doesn't consist of -a reliant of an emigrant, reliant such vulnerable accompanies that emigrant or migrates consequently for the reason of allowing that emigrant in the nation to which that emigrant has legally emigrated; or any person who has lived outside any nation at any duration after completing 18 years of age, for not fewer than 3 years or the companion or offspring of such person (Cummings, 2016). Some reasons make emigrants move and they are categorized into push and pull factors.

1. A factor that pulls emigrants to a certain locality:

- Inadequacy of employment or entrepreneurial favorable circumstances;
- Inadequacy of favorable educational circumstances;
- Inadequate political or religious privileges;
- Intimidation of arrest or penalty;
- Affliction or intolerance based on religion, race, sexual orientation, or gender;
- Incapability to find a companion for marriage;

- Inadequate independence to accept beliefs of divinity, or to choose no belief in the divinity;
- Deficiency of farmland; tough to opening up virgin farms;
- Overwhelming legal or political circumstances;
- Striving or a declining economy;
- Military outline, warfare, or crime;
- Hunger or dryness;
- Rivalries with other cultural associations;
- Banishing by armed force or pressure; and
- Congestion.

2. Factors that push emigrants to other localities:

- Conducive letters from relatives or informants who have previously moved; conglomerate migration;
- Better favorable circumstances for getting land for self and children and inexpensive farmland buying;
- Speedy wealth (as in mineral-related areas);
- More job favorable circumstances and Assurance of better salary;
- Prepaid travel (as from siblings);
- Better well-being programs and better education;
- Construction of certain cultural or religious societies;
- Constructing a new state historically;
- Unite siblings who have previously moved; chain migration;
- Political independence and Cultural favorable opportunities;
- Bigger opportunity to find a companion;
- Friendly climate; and
- Permissive to across the outer limit.

3. **Emigrants due to Economy:** Economic migration is the migration of people from one nation to another to an advantage from bigger economic favorable circumstances. It is frequently presumed that such movement is majorly from fewer economically advanced nations to the more economically advanced nations and from formerly protectorates to the nation, that was the stately power.
4. **Emigrants from Political Reasons:** Abundant people are compelled to move due to war, civilian war or nation approaches

that segregate against specific groups of nationals or people who obstruct those in power. These people are not capable to go back home due to anticipations of being tormented and are absurd to receive any preservation from their government.

5. **Emigrants due to the Environment:** Environmental migrants are people who are compelled to move from or run away their home locality because of rapid or long-term transitions to their resident environments, which contrarily affect their welfare or subsistence.

6. **Family Reunion Emigrants:** This type of migration attribute to a position in which relatives come to join one of its representatives, who is legally citizen in another nation. This majorly involves a married local colleague (in nations where such supplies exist), companions, or not presently wedded or, reliant siblings and fully grown and aged reliant members of the family.

5

GLOBAL DISTRIBUTION OF THE HUMAN POPULATION

CONTENTS

5.1. INTRODUCTION

The distribution of the population is the pattern of place in which people live. The population distribution in the world is imbalanced and is categorized into two: sparsely and densely populated. Localities that are sparsely populated contain fewer people. Localities that are densely populated contain bountiful people. Sparsely populated localities have proved to be difficult places to live. Usually, these are places with inhospitable environments like Antarctica. Densely populated places are hospitable. Defining population diversity in the world as the elements of population distribution on a continuum of similitude to heterogeneity concerning a unit or more variables, standards are characterized with variance-based explanations that qualitative data can be applied on. Population distribution goes hand in hand with the density of the population which is the measurement of the people's number living in a locality computed by dividing the people's number by the shown area as people's number per kilometer square. The distribution of the population cannot exist without population density.

5.2. FACTORS PROMOTING DIVERSITY IN DISTRIBUTION OF THE POPULATION

The fundamental factors resolving the distribution of the population are natural and also human.

5.2.1. Natural Factors Affecting the Distribution of the Population

Natural factors include:

- **Climate:** The climatic conditions of a place greatly influence the way the population is distributed for the areas that have heavy rainfall tend to have very high population and the dry areas extremely sparsely populated since the sand doesn't support vegetation unlike in areas for example in the tropics where the populates can get involved in farming activities. Likewise, where temperatures are on extremes the population is sparse unlike where the temperature is moderate and less strong winds the population is always on the uptrend.
- **Landforms:** Being a feature that is natural or artificial on the earth's surface typical landforms consists of plateaus, hills,

canyons, valleys, and mountains, including shoreline features for example; seas, peninsula, lake and bays as well as engulfed features like, volcanoes, mid-ocean ridges, and the great basins. Where these appear, people always tend to populate around them. In most cases, these areas are always highly super densely populated.

- Topography: This is how the physical features and or physical are arranged. Some areas are populated scatter due to the physical features it is like for people who are not feeling safe to live in such areas. This explains why the leeward sides of most mountains are poorly populated for example Nanyuki on Mount Kenya as compared to Meru which is on the windward side of the same amount.
- Soil: The areas with soils that favor agriculture are densely populated and not only that also these areas have good rainfall to support the agricultural activities unlike the areas with so much and also no reliable rainfall and also poor soils.
- Energy and Mineral Resources: People always want to be near the resources or within the vicinity of the resources so in most cases it found that these areas are highly populated but this depends on the kind of energy being produced if poisonous or its in away affecting the health then populates tend to move far from them.
- Proximity to like sea coast distancing, harbors that occur naturally, rivers or canals that are navigable: For the communities which love to be near the water bodies and also involved in fishing activities tend to crowd around these water bodies.

Coastal regions, areas that are home to a large and growing proportion of the world's population, are undergoing environmental decline. The problem is particularly acute in developing countries. The reasons for environmental decline are complex, but population factors play a significant role. Today, approximately 3 billion people — about half of the world's population — live within 200 kilometers of a coastline. By 2025, that figure is likely to double. The high concentration of people in coastal regions has produced many economic benefits, including improved transportation links, industrial and urban development, revenue from tourism, and food production. But the combined effects of booming population growth and economic and technological development are threatening the ecosystems that provide these economic benefits. Unless governments and users of coastal resources take action, population pressures and the associated levels of economic activity will further degrade many coastal habitats.

5.2.2. Human Factors Affecting Population Geography

These include:

- **Cultural Factors:** The cultures of different communities greatly influence their settling and also their distribution, for example, the Maasai community of Kenya settled on the plains where the land was vast to accommodate their large herds of cattle.
- **Political Boundaries:** Every community back to their origin they had a form of the political administration which is still in existence today. These boundaries make some communities not to live in certain areas inhabited by other communities seen as a foe or something of the sought. This makes some areas to remain sparsely populated because of differences in the administration of communities.
- **Controls on Migration:** Sometimes regulations are posed by the leaders of some communities not to migrate to some areas that are susceptible to the eruption of a mountain, disease, etc.
- **Trade:** The areas where people meet for the trading activities are always densely populated since people of all countries and communities gather to exchange goods and this brings cohesion most of the time's intermarriage between the communities.
- **Government Policies:** The government in place may come up with restrictions and regulations prohibiting or encouraging its citizens to move or not move to a certain area. This is affecting the population distribution so much.
- **Types of Economic Activities:** For an instance of the nomads, who don't have a permanent dwelling, always looking for green pastures, this means that the population distribution will have to remain sparse. Also, if a person moves to an area where the original activities were herding and maybe he is a vegetarian then this person will be stopped because he has brought competition.
- **Technology:** Not excluding the farming, type, and facilities involved in transportation, social movements. Michael Kremer combined two basic concepts to explain the greater than exponential population growth in human populations over the

last million years.

The first concept is that more people means more ideas. A larger population will generate more ideas to feed technological progress. The second concept is that, in a Malthusian world, population is constrained by income, with income a function of technology. Population can only increase if there is technological progress, with any increase in income generated by technological progress rapidly consumed by population growth.

When you combine these two concepts, a larger population generates more ideas, which in turn eases the constraint on additional population growth, which further accelerates the production of ideas. The result is population growth being in proportion to the population size. The following diagram illustrates the feedback loop.

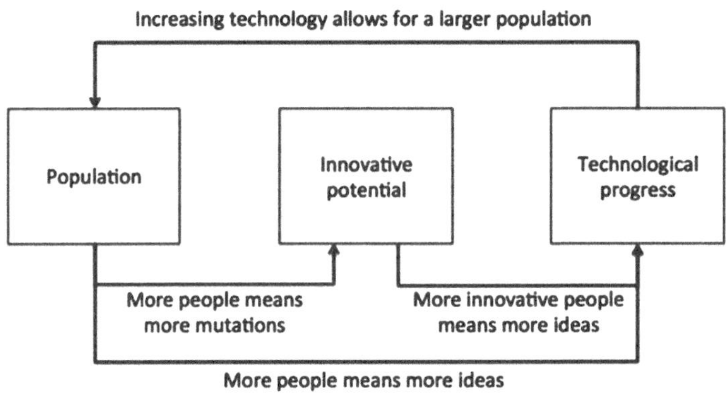

Figure 5.1. Population growth proportion and innovative potential.

Source: https://jasoncollins.blog/2013/07/02/population-technological-progress-and-the-evolution-of-innovative-potential/

Adversity in physical conditions and means of organization is lacking due to insufficient opportunities and but not forgetting demographic factors, for example, natural increment changes and livelihood have been majorly responsible discourager in dwelling in specific areas. Climatic conditions are maybe the majorly essential of all the geographic influences on the distribution of the population. Except, demographic, economic, political, historical factors all of them being placed under social factors, physical factors affect the distribution of the population.

Taking, for instance, Cambodia, civil unrest in the past had made a huge number of Cambodians to move from one province to another. Nevertheless, a large group of populates is directly involved in agriculture and seldom shift their homes in cool times, there have been migrations of the populates majorly for economic reasons. In Cambodia, there is a clear pattern in the distribution of human settlements. For example, most of the populate of Cambodia are settled in quite lasting villages close to the most streams of water in the Tonle Sap Basin-Mekong plain region. In the northeastern provinces of Kratie, MondulKiri, Stung Treng, and Ratanak Kiri major places where highland tribes of Cambodia are found. They majorly live in dispersed transitory villages that are forsaken when the cultivated land in the neighborhood is extremely weak to support plant life. Only a few hundred inhabitants are in these fields. The entire region is occupied by Cham people who are majorly Muslims involved majorly in growing vegetable and fishing. These villages are majorly positioned on or near the banks of streamlines (Bashford, 2014).

5.3. GLOBAL AGE STRUCTURE AND DEMOGRAPHIC TRANSITION

The population age structure is an essential factor in population changes. Structure of age is the portions of differing age division in the population. Models that assimilate the age system permit exceptionable forecasting of population growth, besides, the capability to partner this growth with the height of economic advancement in a locality. Countries with accelerated production have a pyramidal configuration in their age systems diagrams, showing a preponderance of youthful human beings, a lot of whom are of conceptive age. This trend is most often noticed in impoverished nations where human beings don't live to oldness considering below optimal living circumstances, and the birth rate is high. Structures of the age of areas with lackadaisical progress, involving developed nations, for instance, the United States, still have a pyramidal makeup, but with numerous fewer young and conceptive-aged individuals and a bigger percentage of aged individuals. Other advanced nations, for instance, Italy, have zero growth in population. The structure of the age of these populations is more pyramidal, with an even bigger proportion of intermediate-aged and older individuals. The real rates of growth in diverse nations are presented with the highest rates inclined to be in the fewer economically advanced nations of Asia and Africa. All nations' populaces can be categorized into 25% of considerable age system typologies grounded on their advancement over the transitions in demographics, that is the decennially-lengthy alteration that major nations have pursued from huge deaths and rates potency to high expectancies in life and later, to the tinier extent of the family. The four age system typologies describe the population dividend that constitutes juveniles and adults that are young below 30 years of age to the dividend of older adults aged 60 and above. Nations with structures of an extremely young age are those in that 67% or extra of the populace is below thirty years of age. Those with a young system of age have started the demographic change but also have over 60% of their younger populace over 30 years of age. A changing structure of age happens when betwixt 45%–60% of the populace constitutes young below years of age. Nations with age system that is mature have rates of fertility at the reinstatement level of 2.1 or fewer children in a woman; below 45% of the nation's populace is 30 years or below, although up to 25% of the populace constitutes elderly adults aged 60 and higher (Bashford, 2014).

5.3.1. Development and Structure of Age

Anthropology is an essential tool for advancement. Youth are supposed to be one of the major commercial advantages for a nation or a locality. A decreasing percentage of toddlers and a concurrent acceleration in the youth dividend and functioning-age adult populace decreases reliance rates and unveils a door of chances for commercial advancement as structures of age grownups and a bigger proportion of the centers of populace (Bloom et al., 2003). This "dividend of demography" can be derived across basic rules, which establishes that when bigger currents of young population come in the market of labor, they are competently taught and jobs remain for them.

5.3.2. Conflict and Age Structure

Between 1970 to 2007, 80% of all new communal rivalries happened in nations with not less than 60% of the immature than 30.5 years of age population the chances of involvement rivalry is biggest among nations with age structures being "very young," where younger population under age 30 is up to 77%. From 2000 to 2007, 67% of all fresh epidemic of communal rivalry happened in nations with super young systems of age: Guinea, Haiti, Nigeria, and Mali, and Cote d Ivoire. PAI's (Population Action International) discovery are enforced by experimental analysis by Henrik Urdal at the International Peace Research Institute, who shown that after regulating for level of advancement administration typology, absolute size of population and last epidemics of rivalry, nations with a big youth growth were 150% more probable than those with major equitable systems of age to feel communal rivalry in the past 50% of the twentieth century. The cause is specifically tenacious for nations with progressing high fertility rates (Lutz et al., 2017). While the connection between age architecture and imbalance is not one of easy matter and effect, the arrangement is persistent.

5.3.3. Governance and Age Structure

The structure of age is also connected to equality. Betwixt 1970 to 2007, thirteen portions of nations with extremely young systems of age were classed as total democracies, correlated to 81% of nations with structures of age that are mature. Other parts of administration, for instance, independence of speech and governmental liberties, show akin correlation. Nations in which above 60% of the community is young than thirty years by age are primly presumable to come across constraints on political independence and national independencies and involvement in dishonesty, feeble corporate size, and administrative quality, all affairs that influence Uganda, Yemen, and Haiti.

5.4. GLOBAL DEMOGRAPHIC TRANSITION

Both the advancing and impoverished nations are experiencing consequential dynamics in their structures of age with possible essential connotations for economic progress. The measure of the transitions depends, but importantly each nation in the globe has accomplished or will accomplish a sub sequentially gain in the population share, of intensification in the industrious ages. On their facets, this advancement has an explicit, agreeable cause on per capita income. Accustomed hooked productivity per laborer, labor effort cooperation ratio, and percentage of unemployment, an increase in the portion of the industrious-age populace will edge, as an issue of easy arithmetic, to acceleration in productivity per capita—the early dividends in demography.

The early demographic allowance typological pasts for a decade, but it is inherently momentary in the description. As the aging populace starts to control demographical tendencies, the division of the populace in the decrease of industrious ages increases. The early share will change to gloomy as the populace progress outstands the growth of labor force. Finally, the population dividend in the industrious ages may be no better than initially when the bonus duration started.

The alike circumstances of demography that yield termination to the early dividend, nevertheless, may direct to a next demographic bonus. A major economic obstacle for populations that are aging is to cater for aged usage for elderly people who commonly have subsequently decreased work earnings. Some communities are arduous to fix this problem by relying on relocation schemes—one communal program or familial backing arrangements. Other communities are answering by accelerating their preserving ratios and accruing bigger wealth physically. It's in this last call that exploration for extra fast-commercial advancement is reinforced in 2001, Ahlburg and Jensen; furthermore, the second bonus is in the momentary in the description. Ageing Population produces a "long-lasting" progression in income and therefore, income per capita (Bashford, 2014).

The revived attraction in the large-scale economic results of transition in population can be foot printed to recent confirmation that emerges in two faces. One, a number of experimental investigations thought on cumulative level board information terminate that demographic aspects have a powerful, statistically meaningful cause on cumulative redeeming ratios. In contradiction, past investigations grounded on smaller durations succession got meaningless statistical backing for powerful causes of demography. Next, comprehensive anamnesis of the wonderful economies of South-Eastern and Eastern part of Asia administer constraining and dependable proof that the bonus of demography was a critical donor to that locality's commercial progress (Bloom, Williamson, 1998) and usage of econometric investigation to finalize that approximately 33% of East and East of Southern Asia's acceleration income per capita was because of the bonus in demography. In 2000 Mason utilizes gain computing approaches to approximate that the bonus regarded responsible for 25% of the region's commercial advancement. Admitting to the system of age variables has the repetitive ability and can "expound" (statistically sense) a major percentage of economic progress, the communion between variables in demography and the recession is not regulatable. Comparatively, the recession results from demographic transitions are ruled contingent.

Asian Tigers' knowledge brings about very bright proof in reinforcing this perception. A lucrative export-adapted progress plan yielded additionally sufficient jobs to swallow the fast progressing workforce. Resistant large-scale economic surroundings—until the catastrophe in finances of the last minute 1990s came—were appealing to finance. As-you-go pay large scale allowance programs that underrate redemption and work enticements have refrained. These and additional rules functioned in recital with the demographic transition to yield big ratios of investment and saving, fast progression in business, and iconic commercial advancement. In the absenteeism of supporting commercial rules, the dividend of demography can't be numbered on to yield profitable commercial outcomes. Approximates of the early and second dividends of demography are made for all nations of the globe in 2005; United Nations (UN) World Population Prospects yields approximate and overhangs. Nation approximates are cumulated into applicable nation associations and used in comparison to the knowledge of the advanced and the impoverishing globe and to contradict essential variations in the progressing globe (Bashford, 2014).

5.4.1. Demographic Transition Characteristics

This part analyzes the transformation of major demographic features back from 1950 and evaluates how the demographic scenery is likely to transit through 2050. It describes the wide differentially of movements through regions and associations of nations with analogous levels of economic advancement, showing how the transition in mortality, fertility, migration, and expectancy of life expectancy resolve history and expected trends at the worldwide level and how they are growing the population centers around the globe.

5.5. GLOBAL FERTILITY RATE

The rate of fertility determines the number of deliveries that take place per 1,000 women of ages between 15 and 44 in a specific year or live births number per 1000 of the population; birth rate indicates to this extent within specific age groups. The rate of fertility can be either high or low. Both of them have pros and cons as we shall look afterward. 4.5 to 7 children fertility rates were common per woman in the pre-modern era. During that, time infant mortality was very elevated which kept the growth of population depressed. As health progresses and the population deaths decreases, we typically saw catapulted growth of population. As the rate of fertility decreased and comes near two children per woman a sudden growth of population comes to a not. The global average rate of fertility today per woman is just below 2.5 children. The global rate of fertility has halved for overly the last 50 years. The children number per woman decreases varies periodically throughout the modernization of societies (Cincotta and Gorenflo, 2011). Seeing the fertility rate globally has halved for the past 50 years are supposed to ask us a question or rather some questions for example: What has caused this low fertility rate among women globally with the improvement in technology in medicine and other health-related fields?, What are the effects to the global economy? What are the possible cures to these deteriorating rates of in the world?

5.5.1. Causes of Low Fertility Rates Globally

In most industrialized countries globally, fertility now stands at heights below those required for replacement of the population. The availability increased and contraceptive effectiveness has enabled women to prevent unwanted deliveries. Moreover, involvement with family life quality, child survival improvement, the labor force accommodating more women, and unreliability of children for old-age security have combined in the reduction of the number of children women desire. Rates of marriage are decelerating, and later age marriages or disruption by divorce are the factors that limit the number of children delivered. The outcome of this is the decline in prolonged fertility in globally industrially countries, the number of children and young adults is decelerating and that of elderly growing. On top of that, the amount of the world population residing in developed countries has dropped from 35% at the beginning of the twentieth century to 24% and is expected to go down more to 17% between 2000 and 2025. These trends will have thoughtful social, political, and economic implications. In specificity, aging numbers of old people will conceive a demand for expansion of social services and the social security costs will be borne by a decreasing labor force. Large-scale displacement from growing countries, where the population of the young adult will go on expanding, can be expected and is a probable spring of political misunderstandings in the industrial countries of haven. The USSR, France, and several East-European countries have embraced pro-natalist population policies--including tax advance to families with children, maternity leaves that are lengthy, daycare facilities at workplace, and in housing allocation preference to young couples--to avert decreasing rates of fertility, but it is not clear whether the impulse of such regulations will be long-lasting (Cincotta and Gorenflo, 2011).

This being a matter of concern to all of us, it has been found that the women empowerment, the acceleration of well-being and status of children's, economic, and technological advancements, changing norms, and family planning privileges matter for the total fertility rate (TFR) reduction.

Women empowerment-the education level in society - specifically women—is one of the major influential indicators for the children's number the families can have. In that a question of costs of opportunity and changes in education influence the choice of having a child, the positive impulse on better health and fewer children mortality as a result of better education of mothers, social norms also change because it's made possible by better education and not leaving behind the contraceptives. Nevertheless, the constitutions of some nations of the world have mostly accommodated women in deciding even on the matters of childbirths.

Decreased men infertility though women are healthy and educated unlike in the mid-nineteenth century, men were strong and huge and the diseases that attack men's fertility at those times are not common lie these days.

Female education is especially important. Research consistently shows that women who are empowered through education tend to have fewer children and have them later. If and when they do become mothers, they tend to be healthier and raise healthier children, who then also stay in school longer. They earn more money with which to support their families, and contribute more to their communities' economic growth. Indeed, educating girls can transform whole communities.

One of the most effective ways to lower population growth and reduce poverty is to provide adequate education for both girls and boys. Countries in which more children are enrolled in school—even at the primary level—tend to have strikingly lower fertility rates (Figure 5.2.)

Rank	Country	Primary School Enrollment	Total Fertility Rate
		Percent	Number of children per woman
1	Japan	100.0	1.3
2	Spain	99.8	1.5
3	Iran	99.7	1.8
4	Georgia	99.6	1.6
5	United Kingdom	99.6	1.9
...			
181	Equitorial Guinea	53.5	5.3
182	Guinea-Bissau	52.1	5.7
183	Djibouti	40.1	3.9
184	Sudan	39.2	4.2

185	Eritrea	35.7	4.6
Note: Rankings are based on a list of 185 countries for which primary enroll-ment data are avail-able.			

Figure 5.2. Primary School Enrollment and Total Fertility Rates for Selected Countries, Latest Year.

Source: EPI from UNESCO

5.5.2. Causes of High Fertility Rate Globally

The Total Fertility Rates serves as the equate lifetime deliveries per woman shown by the age-certain fertility rates overcoming in one historical mileage.

The causal impact assessment of high fertility is an investigative challenge because virility is, to a more or lesser degree, a preference, that is, it is internal. Fertility covariation with other results—economic, social, and health—may mirror cautiously picked trade-offs rather than a straight forward casual effort of fertility. High fertility has the following implications:

- **The Health of the Child:** The infancy mortality risk and early childhood are preeminent for higher-order deliveries and closely-separated deliveries, and when the age of the mother is over 40.
- **The Health of the Mother:** The risk of maternal mortality risk is preeminent at higher consistencies and younger and older ages. Fertility decline reduction causes the lifetime risk of mother's death just by decelerating the average number of parturient each woman experiences.
- **Schooling of the Child:** Children from large families attain less schooling and repeatedly deliver cohorts—a societal feature of high fertilities—removes from the quality of schooling by unsaturation of per-pupil expenditure.

- **Growth of the Economy:** In the long-run, fertility raises production output exogenously. And the correlation between the growth of the population and the growth of the economy has already been negated since the 1980s.
- **The Dividend of Demography:** Economic growth is assisted by a family decline through favorable changes that are favorable to the age-structure—the "demographic dividend" concentration which is larger of the population in the working ages, therefore accelerating productivity per capita. Economic growth in East Asia and Latin America in the period since 1960 has been contributed substantially by the demographic dividend.
- **Natural Environment:** The cause of looming shortages of freshwater in many countries is a direct and proximate of high fertility and resulting population growth. Global warming has been contributed by population growth—the addition may be as much as a third—and reduction of fertility through expansive family planning.

A major development from the more modern job on the joint between development and population is the importance of being certain about the ways through which the joints work. At the major level, the impingement of high fertility on other results could be directed through the population size (natural environment implications), population growth rate (budgets implications), or population distribution of the age (economic productivity implications).

High fertility is influenced by the following factors:

- **High Demand for Children:** The children demand is high in most of the high fertility rates countries especially in the West and Central Africa.
- **Unmet Family Planning Needs:** Most of the high fertility countries unmet the medium to high levels of family planning needs—the pervasiveness typically ranges from a fifth to a third of women who are married.
- **First Union Age:** In most high fertility rates countries, society's age at first union is relatively young (on average 20 or less age). Delay would contribute to fertility decline over several years, and it would have other socioeconomic benefits and health and socio-economic benefits.

- **Mortality:** The most powerful stimulant of fertility decline is improved child survival. In comparison, mortality increases due to the HIV-pandemic is having a low general impact on fertility rates and growth of the population.
- **Education:** One of the most powerful tools in stemming population growth is education. A more-educated workforce also makes poverty eradication and economic growth easier to achieve. Non access to education, specifically reproductive health education may pose a threat to increase the population.
- **Income:** By comparison, income is a relatively poor indicator of the decline in fertility, the net income for education, food, household needs is essential for productivity and development. Economic performance of the family is intertwined with poverty and not itself a challenge to fertility decline.
- **Contraception Obstacles:** Non-access obstacles (psychic, social, cultural) appear to be major in some settings but are not well measured.
- **Services of Family Planning:** The data on access challenges is less ambiguous: in divergent settings expanded the availability of family planning services has had an impulse on fertility, a net reduction in fertility being 10%–25% typically.
- **Religiosity:** Various scholars have seen this theory as widely about religious arrangements and involvement. It includes ritualistic experiential, ideological, creedal, intellectual, communal, consequential, moral, doctrinal, and cultural dimensions in culture. Religion sociologists have viewed that the beliefs of people, belonging sense, and behavior at times is not coherent with an individual's real religious beliefs because there is much change in how religious or not one can be. Multiple challenges persist in measuring religiosity majorly promoting or degrading contraceptive usage.

From a cultural standpoint, religion can have a profound effect on family planning. Many religions promote large families as a way to further the religion or to glorify a higher power. For example, Orthodox Judaism encourages large families in order to perpetuate Judaism. Roman Catholicism promotes large families for the same reason, and forbids the use of any "artificial" means of birth control. Devout followers of a religion with such values often have large families even in the face of other factors, such as economic ones. This can be seen in countries like Israel (Judaism) and Brazil (Catholicism), which have high percentages of religious followers in their populations. Both countries have high birth rates and high population growth rates.

Figure 5.3. Church's opposition to contraception's.

Source: https://www.rappler.com/move-ph/issues/gender-issues/87725-divorce-philippines-survey

5.6. GLOBAL MORTALITY RATE

The rate of mortality, or rate of death, is a measure of death number (generally, or because of a particular reason) in a specific population, narrowed to the population size, per time unit. The rate of mortality is generally shown in units of deaths in 1000 people per annum.

The rate of death globally is influenced by various factors; these factors contribute directly or indirectly to the increase or decrease of mortality rates. Among them are diseases, injuries, wars, intentional or unintentional food poisoning, unsafe WASH services, air pollution, tobacco usage, and health (Lutz et al., 2017).

5.6.1. Mortality Due to Health

According to World Health Organization, "Health is very vital to development: it is a contributor to, a precondition for, and also an indicator and results of progression in maintainable development" (WHO, 2012). Over the past few decades, there has been substantial advancement globally in advancing various factors that affect the health of the population, including access to safe waters, nutrition, sanitation improvement, and living standards in general. The millennium development goals (MDGs) paid so much interest in major health aspects and population mortality, including child health and maternal health as well as tuberculosis, malaria, and HIV/AIDS.

It is a challenge to know for sure whether the MDGs just drew the scrutiny to the advancement being made in these areas or either the Goals helped to speed the acceleration of that advancement. Whichever the case, the exceptional accomplishments of the MDGs era must be maintained and emphasized if the world is to accomplish the new and more aggressive objectives of the Goals of Sustainability in Development as endorsed by the UN General Assembly.

The SDGs invite the following achievements by 2030:

- Reduction of the ratio of maternal mortality to fewer than seventy per 100,000 live deliveries;
- Close avertible immortality of infants and below the age of five years children, with total nations aspiring to neonatally decrease mortality to below 12 per 1,000 live deliveries and mortality of under-five to slightly low as 25 per 1,000 live deliveries. End the endemic of malaria, tuberculosis, AIDS, and undervalued equatorial diseases and fight diseases that are waterborne, hepatitis, and alternative easily contractible infections;
- Decrease by third, immature deaths from non-contractible diseases (NCDs). By avoidance and medication, in addition to encouraging psychological well-being and health;

Improvement of health services and providence of worthwhile and high-impulse interventions that talk about the needs of women and infants across the care of the continuum, with attention on care during the time of delivery, remain important as a way of saving lives (Lutz et al., 2017).

Pneumonia, malaria, and diarrhea remain the first effects of death among children below age five. 45% of under-five deaths happen in the first month of life, or period of neonatal, when deaths are influenced by factors related to pregnancy, along with intrapartum and preterm challenges (between 48%–61% range of neonatal deaths throughout the regions) and infections like meningitis, sepsis, pneumonia, and meningitis (between 7%–24%). Within the second across twelfth months of life, or period of post-neonatal, deaths from the dominant contagious diseases (for instance, diarrhea, pneumonia, malaria, AIDS, meningitis, and measles) still report nearly 70% of them all in sub-Saharan Africa.

More than 50% of all deaths in 2016 in low-income nations where effected by conditions of "Group I," as so-called, that include contractible diseases, causes of maternal, pregnancy conditions that arise during and childbirth, and deficiency in nutrition. By contradiction, high-income countries only less than 7% of deaths were due to such effects. Decreased breathing diseases were among the prominent death effectors throughout all categories of income.

NCDs effected 71% of death globally, that ranged from 37% in small-income nations to 88% in large-income nations. Nine the ten largest effects of mortality in large-income nations were on convertible diseases. In the provision of unadulterated death numbers, however, 78% of world NCD deaths happened in small- and intermediate-income nations.

Demographic and health surveys (DHS) on seven fundamental health services in small- and lower-intermediate-income nations were analyzed. These were the seven services: four or more visit antenatal care (ANC) four or more times, tetanus vaccination during pregnancy to be at least one; attendance of skilled birth; vaccination of bacilli (Calmette-Guerin); administering the third dosage of a vaccine consisting diphtheria, pertussis, and tetanus, vaccination of measles; and access to enhanced drinking water in the families (Lutz et al., 2017).

5.6.2. Mortality Due to Air Pollution

Circulating air pollution is responsible for approximated deaths of 4.2 million per year due to heart disease, stroke, chronic respiratory diseases, and lung cancer. About 91% of the population in the world resides in places where the air quality exceeds the limits levels of WHO. Developed and developing countries alike are affected by the circulating air pollution, the world gets more crowded and hotter, dirty emissions pumped out by our engines continually, and the clean fuels or technologies accessibility in the world being half (e.g., lamps, stoves), the very air we inhale is becoming precariously polluted: people who breathe polluted air are nine out of ten; 7 million people every year are killed. The effects of air pollution on health are dangerous—third of deaths are from stroke, heart disease and lung cancer are because of the pollution of air. This is having a commensurate cause to that of tobacco smoking, and exceedingly higher than, say, too much salt eating effects.

5.6.3. Mortality Due to Injury

In 2016 injuries collected 4.9 million lives and (29%) of these mortalities were because of injuries in road traffic which is more than a quarter. Low-income nations had the biggest rate of mortality because of injuries from the traffic of road traffic with 29.4 deaths in a population of 100,000—global rate was 18.8. The traffic Road lacerations existed also between the majorities of 10 death effects, under intermediate- and uppermost-intermediate- nations' income.

5.6.4. Mortality Due to War and Violence

In most countries where there has been war either communal or tribal or even politically motivated violence's very many lives are lost for example a case of Rwanda: violence between the Hutsi and Tutsi communities over the leadership of the countries though it had some political backing where women, men, infants lost their lives and also the post-election violence in Kenya where thousands were displaced others killed and left so many incurable wounds to those left.

5.6.5. Mortality Due to Food Poisoning

Hazardous food accommodating dangerous bacteria, parasites, viruses, or chemical substances, generate more than 200 infections that range from diarrhea to cancer. An approximated 600 million—utmost 10% of world people—fall sick after consuming harmful food and 420 000 die annually, culminating in the deficit of 33 million healthful years of life under 5 years children of age ferry 40% of the food borne infection, every year with 125 000 deaths.

Diarrheal diseases are the majorly common diseases culminating from the usage of corrupt food, causing 550 million people to become sick and deaths of 230, 000 annually. Nutrition, food security, and food safety are inextricably joined. Hazardous food makes an atrocious cycle of disease and malnutrition, specifically affecting the newborns, children that are young, aged, and the ill.

5.6.6. Mortality Due to Tobacco and Substance Usage

Smoking cigarettes contributes to more than 480,000 deaths annually in the United States. This is closely one in five deaths. Tobacco and substance usage smoking contributes to more deaths annually than the following causes put together:

- Illegal drug use;
- Human Immunodeficiency Virus (HIV);
- Alcohol use;
- Fire arm-related incidents; and
- Motor vehicle injuries.

Exceeding ten times as many United States nationals have died immaturely due to a smoking cigar than as many have fallen in all the combats battled by the United States. 9 out of 10 of all deaths are generated by smoking. Women in the majority die because of cancer of lung cancer annually more than they die from the cancer of the breast. Smoking generates deep-rooted intrusive pulmonary disease to about 80% of all deaths. Smoking Cigarettes accelerates the risk of mortality from total causatives in women and men. The risk of death arising from cigarette smoking has advanced in the past 50 years in the United States.

5.7. THE MIGRATION AROUND THE GLOBE

Globular migration administration can be characterized as the norms and institutional structures that regulate and facilitate nations and other actors' responses migration. Its primary purpose is to ensure that states work collectively in ways that make them better able to fulfill their objectives than they would be acting alone. Reflecting wider trends in inter-boundary inter-joining, there has been human flexibility across global borders. Between 1970 and 2017, the number of global migrants accelerated from 82 million to 244 million, constituting 3% of the population of the world. Above 60% of all global immigrants live in Europe are about 78 million or Asia about 80 million. North American presented the number three of global immigrants of about 58 million, followed by Africa with about 25 million, Caribbean, and America Latin with about 10 million and 8 million by Oceania. Two-thirds of all global immigrants in 2017 were existing in just nameable nations. The biggest global immigrants number lived in the U.S.A., Germany, Russian Federation, and Saudi Arabia, accommodated the numbers two, three, and four largest number of immigrants globally (each approximately 12 million), pursued by the United Kingdom and North Ireland (approximately 9 million). In the year 2016, the aggregate computation of aliens and shelter finders in the globe was assessed at approximately 26 million. Turkey accommodated the biggest population of immigrants globally, with about 3.1 million aliens and shelter finders, pursued by Jordan with 2.9 million immigrants, Palestine state (2.2 million) and Lebanon at (1.6 million).

The antiquity decades have also observed a misrepresentation of the worldwide migration map with emigrants from an accelerating different bunch of non-European-origin nations focusing in a decreasing pool of major terminal countries (Ness, 2014). Yet in disagreement to other policy areas which include inter boundary migrations, a systematic UN (Figure 5.4.).

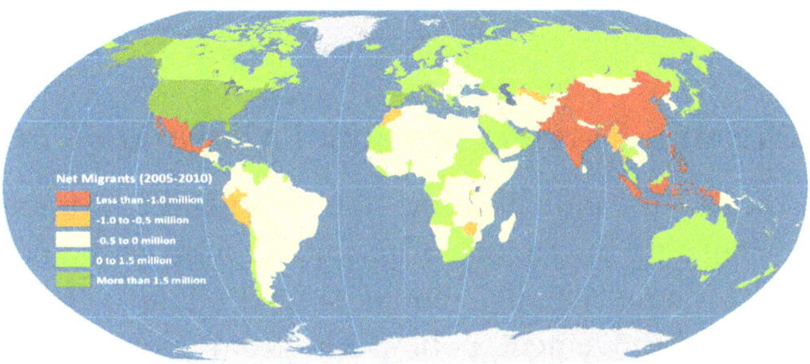

Figure 5.4. The Global net Migration.

Source: UNEP (2012): The UNEP Environmental Data Explorer, as compiled from World Population Prospects, the 2010 Revision (WPP2010), United Nations Population Division. United Nations Environment Programme. http://geodata.grid.unep.ch.

The multilateral administration groundwork has been lackadaisical to emerge. The major pressure on the progress of global administration of migration has been declared view that acceleration in worldwide governance on this issue includes shrinkage of the nation's sovereignty. Global movement administration shows a disintegrated drapery spread of institutions escalated across policy areas and levels of administration and it has emerged constantly over a long duration (Betts, 2011). The disintegration of the dissolution of administration into parts brings with it a range of merits and demerits sometimes facilitate and sometimes necessitate cooperation (Biermann et al., 2009).

In 1949 the Relief of United Nations and Palestine Refugees Works Agency in the Near East (UNRWA) was made to carry out a forthright ad and works schedule for refugees in Palestine. In 1950 the Office of the United Nations High Commissioner for Refugees (UNHCR) was entrenched to secure and find explanation formerly Europe's refugees were given bureaucratic authority for the newly made 1951 Convention on the Refugee Status, which explained who is an alien and the privileges to which such people would be designated. However, a small institution known as the Provisional Intergovernmental Committee for the Migration of Migrants from Europe (PICMME), next to becoming ICM in 1980 and 1989 as IOM, was also made in 1951 to backing logistics describing the migration of populations that are displaced although the first UN World Population Conference assembled in 1954 in Rome and it was majorly a professional meeting, bringing together knowledgeable experts (Ness, 2014).

5.7.1. Benefits and Barriers to Global Migration Governance

There are merits to embellishing the worldwide administration of movements. Managing migrations of people through global borders cannot be attained over single nations' actions singly; slightly, the advancement and application of movement policy aim from global collaboration in super-scribing the compound lead foot and development of movements. By description, global movements include at least two nations—ancestry and terminal—and progressively affect many other nations that aid as movement points, talent competitors, corporates in fighting planned crime and migration of terrorists, and partners in the financial system globally that moves costs. Furthermore, the movement also includes non-governmental players that intersect with states and every other in managing people's migration. Some of these have explicit, approved duties (e.g., multinational collaborations, trade unions, labor enrollment agencies, and humanitarian aid organizations) in as much as others are involved in criminal activities, for instance, migrant smuggling and human trafficking. Moreover, global movements intersect with other international issues, counting trade, security, development, conflict resolution, and change in environment, calamity risk decrease, humanitarian action, and rights of humans. But, unlike these other localities, achievement to establish worldwide administration arrangement to react to actual and emerging obstacles have delayed in the movement locality.

In the confrontation of worldwide collaboration and strategy obstacles, an excellent productive arrangement of worldwide movement's administration can advance corporate answers and make chances for mutual assistance. The analogous arrangement can bring nations in unison to deliberate on issues of collective responsibility; analyze common ambitions and approaches; make the area for literature and discernment; and permit nations to correspond and collaborate, in addition, advancement and application of structure, processes, and actions. Worldwide norms, including guides principles and rules, whether binding legally or not, base criterion against which nations behavior can be evaluated. The welfare arising from the worldwide administration of movements should also be based on the to which analogous schematic administration and any development to it should necessarily be based on the identification and welcome that immigrants, like anyone else, are designated to absolute rights. In his eventual statement, Peter Sutherland -Special Representative of the secretary-general on international migration (SRSG warned that worldwide movement administration cannot be attained by authorization: Advancement is feasible to articulate on the entanglement of those straight afflicted and those accountable for approach application, while reaching concession on accepted least, assumptions, and theories that should administer throughout the board. The recently brings monotony for international collaboration argued on clearly enunciated bilateral expectations and duty, and for immigrants, whose right prerequisite be guarded everywhere they tend to move in the globe (Ness, 2014).

5.7.2. Impacts of Global Migration

1. **Economically Related Impacts:** In 2015, migrants added $6.4 -$6.9 trillion which is approximately 9% of the global gross domestic product (McKinsey Global Institute, 2016). Migrants' aspiration and escorting devotion to upgrade their sustenance are two of their major characteristics. Cities suggest more opportunities and best income, shut the skill gaps, and in some cases necessitates an alternative to the cheap workforce. Likewise, cities bring a wide customer base for immigrant entrepreneurs to sell their goods and services. The talent and labor insufficiency in cities also attribute majorly to healthy competition for trained entities in these communities. Migration's aftermaths on labor markets rely on the artistry of migrants and living workers, and the attributes of the terminal economy. The movement could affect the general economy of the accommodating city through cash (Shera and Meyer, 2013).
2. **Impacts on Social Life:** Migrants mostly find another representative of their household and their indigenous or cultural association in cities because the conglomerate movement is

generally an urban incident. The impulse to reside among one's group is conspicuous and answerable for authorizing circles within cities. Circles offer merits of society and social support, preeminent availability of indigenous goods and food, worship centers and other societal institutions. The number of indigenous circles is increasing in the urban localities of advanced countries, leading in intermediate-class migrants and whole neighborhoods where aliens reside forever rather than being on movement.

3. Impacts on Urban Infrastructure and Services:

- **Housing:** With their outrageous housing prices, the world cities of Mumbai, London, New York, Shanghai, Paris are also among the large cities jolted by migration. Cities in Africa Sub-Saharan also suffer from uncompromising housing deficit, causing huge inflation in the housing price of house units. A shortage of economical housing has appeared in people living in slums or crouching Education and employment ascending immigration directly disturbs the opportunity of localities in primary school, and inevitable pushes schools to increasing class numbers and classroom addition. Migrants have the artistry spectrum ranging from incompetent labor to competent workers and have become a major driver for identical artistry skill demand and supply ratios worldwide.
- **Health:** In cities with a momentous population of migrants residing in slums, migrants' residing environment and other social incentives aggravate physical, social, and mental and health risks. Realities coming from new political and social environments and language obstacles place big appeals on the enduring skills of first- and second-generation migrants and their capability to accustom.
- **Water:** The incursion of migrant's localities an accelerating demand for resources of water, pursued by the associated progress in the generation of sewerage which, in turn, makes the demand for treatment of wastewater.
- **Energy:** Migration changes energy consumption and CO_2 emissions significantly. Energy consumption approximates are lower when the city population has advanced due to rural-urban migration than over urban-urban migration and growth of population naturally.

- **Information and Communications Technology:** The increasing agility of information has changed migration discharge and accelerated the chances for huge arrivals squeezed time frame, thus affecting the city of the destination.
- **Waste and Sanitation:** Migration can greatly aggravate the problems of managing sewage given the expansion of the population, but the city cannot consistently meet the demand due to inadequate capacity.
- **Security and Safety:** Some advanced countries regard the movement as a security issue and use it as a justification to instate strict and curbing policies. These policies limit shelter seekers' access to secure countries, and in severe circumstances lead to smuggling migrant and trafficking of humans, or their taking insecure routes or passages. The emerging dissimilarity between the migrant's interests, governments of cities, and countries excessively monitoring the operations of migrants have led to a misaligned attitude on exodus in such nations.

6

POPULATION DIVERSITY IN THE SEVEN CONTINENTS

CONTENTS

6.1. INTRODUCTION

There are seven contents in the world. These continents have differences in terms of population diversities. This chapter was prepared to offer key information about these diversities as discussed below.

6.2. EUROPE POPULATION DIVERSITY

Europe is progressively a locale noted by ethnic diversity and alien-born populaces. Through the ten European Union's countries scrutinized, the average share of immigrants in the populace is 12.2%. Nevertheless, the percentage of immigrants in individual nations spheres from as great as 18.3% in Sweden to as small as 1.6% in Poland. The portion of immigrants is advancing in some nations, largely due to the big number of shelter seekers arriving in Europe in the last year. In Hungary and Sweden, for instance, the percentage of the populace that is alien born advanced by 1.5 proportion points and 1.3 proportions points mutually between 2015 and 2016. At the same time, the percentage of the populace that is alien born has not developed at all in Poland and has decreased by 0.1 proportion points in Spain over the same duration (Figure 6.1).

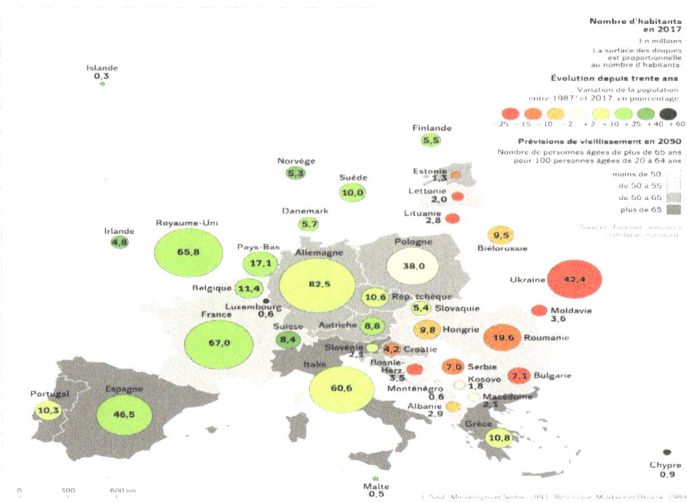

Figure 6.1. Population diversity in Europe.

Source: https://mapsroom.com/pin/1480/.

In this scrutiny, people were asked whether having an accelerating number of people of many different races, cultural associations, and states in their communities makes their nation a better place to reside, a poor place to reside or doesn't make much contrast either way. In no state does a bulk say advancing diversity is certain for their nation. At most, approximately 33% in the UK (33%), Spain (31%), and Sweden (36%) construe advancing ethnic, racial, and country's diversity in favorable terms. More general is the perception that diversity of culture is neither nor a decline in terms of life quality. This is the dominant approach in Germany, France, Spain, and the Netherlands, where advantages say increasing diversity makes a small difference in their life quality. In the meantime, the British public is divided betwixt those who see diversity as a considerable, diverse or non-aspect for their nation. At the same time, more than 50% of the public in Italy (53%) and Greece (63%) admit that advancing diversity makes their nation a poor place to reside. Perceptions of the value of the country's diversity often separate along mental lines. People who self-recognized as being on the conservative side of the mental spectrum are necessary more probable than those on the liberal to say that advancing diversity is damaging to their nation. Space is the biggest in Germany, where those on the left and right are 36% points apart. In other nations, the mental split is also important, containing in Netherlands, Greece and Italy.

Party relationship is also joined to perceptions of diversity culturally. In the United Kingdom, for instance, a total of 64% of people who back the right-wing United Kingdom Independence Party say that a more ethnically nationally, and racially mixed community makes the United Kingdom a poor place to reside. Only 32% of Conservative Party supporters and Party of Labor 19% adherents share those perceptions. In France, 51% of those who feel nearest to the anti-immigrant State Front say acceleration diversity is atrocious for France. Just 34% of adherents of the Republicans and 11% of Socialist Party supporters concur. In Germany 62% of people who have a considerate view of the conservative political party alternative for Germany trust that Germany is worse off due to advancing numbers of people with diverse ethnic-national and racial identities. And in Sweden, 65% of the people that hold an approving view of the anti-immigrant Swedish Democrats say the belief that more mixture is making Sweden a bad place to reside.

The perception of diversity also contrasts grounded on the academic level of the respondent. Less-learned people are more detractive of diversity than more-learned members of the nation. In the Netherlands, 43% of people with a subordinate education or lower say an advancing number of people of many diverse ethnic groups, races, and nationalities make their nation a bad place to reside. Just 22% of those with more than a subordinate education concur. There is an analogous 20-point academic space in perceptions of diversity in the United Kingdom, Spain a 14% gap and a Sweden divide in 13%.

6.2.1. Accent: A Tenacious Necessary of the Country's Identity

The EU has twenty-four legitimate accents and several other locational and minority accents among its twenty-eight representative nations. And accent facility is looked like an essential constituent of nationality throughout Europe. More than 90% of people in all of the countries scrutinized say that to be a real national of their country it is essential to speak the nation's national accent. Such a belief is not cheaply held. Most in all of these nations say it is very essential to be capable to converse in the resident tongues. People on the right of the mental spectrum are more probable than those on the left to say a national accent facility is very key (Wilson, 2018). This space is the biggest in Sweden, the UK, and France. In some nations, people aged 50 and above are more probable than those aged 18–34 to firmly trust that the capability to converse the official language is very essential. This generation gap on accent is 17 points in Spain, 23 proportions points in Sweden, and 18 points in the United Kingdom. Americans also view language facility as key to the country's identity nationality roots of culture.

Most Europeans accept that complying with inherent traditions and customs is also essential in describing the country's identity. Vast bulk in all ten (10) European Union nations expressed the perception that sharing such facets of culture is key to being a real Pole Swede or German or and so forth. But there is less passion for such belief than there is about conversing the national accent. 50% or more of respondents in only 5 nations say sharing traditions and customs are very important. This tenacious belief joining culture and nationality is most favor in Greece (66%), Hungary (68%) especially, 26% of Germans and 36% of Swedes expressed the perception that it is not very key or not essential at all to share inherent traditions and customs to be treated German or Swedish.

There is a proportionately deep mental split over the relationship betwixt devotion to local true and culture nationality. Those on the right are necessarily more probable than those on the left to join the two matters in several nations. In the United Kingdom, this right-left numerical is 30 proportion points. The gap points in France is 29, and 21 points in Poland. Identity of culture is also a biased matter in a few nations. In France, splitting French traditions and customs is most essential to those who feel nearest to the Front of the Nation. But socialists are far less probable to feel tenacious about this. In the United Kingdom, 81% of those who established with UKIP (UK Independence Party) say complying with the culture of the British is very key to being British. Only 44% of labor comforter concurs. People of different generations disagree on the essence of traditions and customs to the identity of the nation. In 80% of the countries scrutinized, those aged 50 and above are more probable than those aged 18–34 to say complying with inherent ethnicity is very essential. This is the case in the United Kingdom (24 proportionate points), Greece (21 points) and France (23 points). Academic history also determines in a person's perception of the joint betwixt national identity and culture. People with less learned are more probable than those with highly learned to trust that tradition and customs are very essential to a nation. These differences in education are 20 points in Spain and France and 19 points in the United Kingdom nationality by birthright.

No European nation gives citizenship grounded directly on the case that a person was delivered in the state's domain. Although, in 60% of European Union countries, bulk say it is very or considerably essential to have been delivered in a nation's outskirts to be treated a real national of that community. The nations where such beliefs are thickest are Poland (42%), Greece (50%), and Italy (42%) and in Hungary, 52% said locale of delivery is very essential). Especially, in Sweden (8%), Germany (13%), Netherlands (16%), and smaller than 20% trust birthplace is a very key constituent of the identity of a nation. In some country's culture plays a big role in such perceptions. Those on the right are much more probable than those on the liberal side to say the place of birth is very essential in Italy (23 points) the United Kingdom (24 points) and Greece (31 proportionate points).

There is also a biased split on nationality according to birthright in few nations. Especially, in the United Kingdom 57% of UKIP advocates, but labor backers only 24%, trust it is very key to be born in the United Kingdom to be treated British. In France, National Front 41% of supporters, but Socialists 21%, say that to be French it is very key that one is delivered in France. And in Spain, 49% backers of People's Party and 42% of Workers' Socialism of

Spanish Party advocates, but just 19% of anti-creation Podemos followers, voice the perception that to be Spanish it is very key to be Spain born. Academics also play a part in public perception of the association betwixt being the inherent born and identity of the nation. In 80% of European Union nations, people less learned are necessarily more probable than those more learned to trust that it is very essential to have been delivered in their country to be treated a real national of that community. These educational differences are 19 points in Poland, 23 points in Spain 24 and proportionate points in the United Kingdom.

6.3. AMERICA POPULATION DIVERSITY

As the United States becomes more culturally and ethnically different, and as industries from Silicon Valley to Wall Street wrestle with how to make labor pool that mirrors these transitioning demographics, Americans have a convoluted, even antagonists, set of perceptions about the impulse of diversity and the best way to attain it. Most say it's a better thing that the nation has a differing populace, but many also say this brings its own set of obstacles. And while having many diverse organizational values, few support the concept of taking race or ethnicity into deliberation in appointing and promotions, conferring to a new Pew Investigative Center estimate (Parrillo, 2015).

When it comes to differences in the societies where they live, most United States are adults; about 66%—containing most of those who live in surrounding with small diversity—are composed with the racial mix in their locale. The most of about 54% of states children should go to regional schools, even if those outcomes in many schools are being less different. Smaller portion of approximately 42% say children should attend schools that are culturally and ethnically combined, even if that says some students go to school external of their regional society. Generally, Black, Hispanic, and White adults are about intermediately probable to say it's better than the United States. The populace is culturally and ethnically balanced, and majorities through these associations say this has had a considerable impulse on United States culture. But dark Americans place more rate than whites and hispanics on organizational diversity and school assimilation (Figure 6.2).

Figure 6.2. Diverse population in America.

Source: https://www.cepal.org/en/news/latin-america-population-will-reach-625-million-inhabitants-2016-according-eclac-estimates.

Assessment on these matters also varies positively along party lines, with Democrats and those who side to the Democratic Party more probable than Republicans and Republican siders to show certain views of the significance and impulse of cultural and ethnic diversity. This is the case even after taking into considering the diversities in the racial combination of the two parties. More than 50% state that racial and cultural diversity is advantageous for the nation.

Most of Americans around 57% say the case that the United States where the populace is made up of people of many different races and cultures are a very advantageous thing for the nation, and another 20% say this is considerably better. Small portions say 5% to 1% considers this is very bad, whereas about 17% say it is neither better nor worse for the nation. Similar portions of 55% whites, 60% Hispanics, and 59% blacks say racial and cultural diversity is very essential for the nation. Blacks and whites with at least a bachelor's degree are more probable than their less learned parties to say it's very essential for the nation to be made up of people of many races and cultures. Approximately 70% of whites with a bachelor's degree or more learned about 69% say this, in comparison with whites 55% with college and 41% of those with a high school diploma or less learned. Between blacks, at least 80% with a bachelor's degree say the nation must be racially and culturally combined; a smaller bulk of blacks with some college was 65% and 44% of those with less learned say similarly. The joint between learning and perception of diversity is a little bright among Hispanics (Parrillo, 2015).

Throughout party lines, most Republicans and Democrats say it's at least considerably better than the United States. The populace is racially and culturally mixed, but Democrats are far more probable to say this is a very essential thing. Between Democrats, whites are more probable than blacks to say diversity is very essential for the nation 67% of Democrats from Hispania say the same. The majority see diversity having a considerable impact on the culture of the nation. 64% of Americans say the case that the United States populace is made up of people of many different races and cultures has a considerable impulse on the culture of the nation; 12% say it has a diverse impulse and 23% of them say doesn't make much disparity. 70% of Hispanics are more probable than whites about 64% and blacks about 58% to say the nation's racial and cultural diversity has a considerable impulse on the nation's culture, though majorities through all three associations say this is the case. This perception is extensive among Democrats, specifically whites from Hispanic. 84% white Democrats and approximately 75% of Democrats from Hispanics say this diversity has a considerable impulse on United States culture, in comparison with of black Democrats 60%. Although they have considerate perceptions of diversity, the majority says it poses obstacles for policymakers.

A considerable proportion (47%) of Americans say having a populace that is made up of people of many different races and cultures makes it difficult for policymakers to answer the nation's challenges; a small share of 7% say it makes it simple for policymakers and almost 50% say it doesn't make great contrast. Whites about 52% are more probable than (42%) Hispanics and (30%) blacks to say the nation's diversity makes it difficult for policymakers to resolve challenges.

Wide backing for organizational diversity, but most say applicants' race and culture shouldn't be an aspect in appointing and promotions. 75% of Americans say it is real (49%) or considerably (26%) essential for industries and organizations to support racial and cultural diversity in their places of work. Blacks are specifically probable to say this is very essential: 67% say this, contrasted with Hispanics 52% and 43% of whites. And the case views about the impact of diversity has on the country in general, Democrats are much more probable than Republicans to say employers need to support racial and cultural diversity in their places of work. These biased contrasts remain when viewing only at those whites. Momentarily most Americans say it's at least considerably essential for industries and organizations to back racial and cultural diversity, only approximately (24%) say that, in addition to their qualifications, a person's race and culture should be looked at in reasoning about appointing and promotions to accelerate diversity. A majority of about 74% say employers should only consider a person's competence when making these decisions, even if it outcomes in less diversity in the working place (Parrillo, 2015).

The perception that employers should only consider a person's competence is extensive among whites is about 78% and (69%) Hispanics; (54%) of blacks parts this perception. Republicans are much more probable than Democrats to say anyone's race and culture shouldn't be a part of appointment and promotion conclusions. Even so, the bulk of both groups hold this perception. Blacks, whites, contrast on the essentiality of students going to a racially and culturally mixed school and going to a school in their local society.

Compulsory integrative programs aimed at integrating schools concluded years ago, but the matter of school segregation remains an issue to argue in many locales of the nation. Currently, (54%) Americans say students should go to school in their region society, even if it means that most schools are not racially and culturally diverse; 42% say students should attend schools that are culturally and racially mixed, even if it translates that some students don't go to school in their regional society.

White and black adults have broad differing perceptions on this. (62%) of whites say students should attend schools in their regional society, even if the majority of schools are less mixed as an outcome. In comparison, (68%) blacks' reason students should go to schools that are culturally and racially mixed, even if they're not in their regional society. Asians and Hispanics are about equally divided on this matter. Perceptions also contrast significantly along party lines, with 73% of Republicans and Democrats 41%—stating students should attend school in their regional society, even if it represents most schools are not racially and culturally diverse. (71%) Black Democrats are much more probable than (52%) white, (56%) Hispanic and (55%) Asian counterparts to say students should go to schools that are culturally and racially mixed, even if its outcome in some students not attending their regional schools.

Approximately 25% of Americans wish their local society was more racially and culturally diverse. 66% of Americans are content with the racial diversion in the society where they live; 24% wish their society was more racially diverse and 8% wish it was barely so. Hispanics and, whites' perceptions about the racial diversion in their society are strikingly persistent unconcerned of how mixed the society is. In this scrutiny, diversity is calculated in conditions of the percentage of a surrounding populace in its biggest racial or ethnic association. The least mixed surrounding is those where 90% or more of the inhabitants are a portion of the biggest racial or cultural association; the most mixed are surroundings where fewer than 50% of the inhabitants are in the biggest group (Parrillo, 2015).

Approximately 70% whites and almost 60% Hispanics—in the greatest and least mixed surroundings say the society where they reside is as racially diverse as they'd love it to be. Among blacks, those in the least mixed surroundings are less content with the racial diversion in their societies. The bigger bulk of whites who reside in surroundings with very small diversity are neighbored by other whites. By difference, for blacks who reside in less mixed locales, their surroundings are less probable to be made up dominantly of black inhabitants. Between black respondents, 63% of those in locales where 80% or more are in the biggest racial or cultural association are in bulk black societies; 34% are in dominantly white surroundings. Vast shares of Hispanics (82%) who reside in less mixed surroundings reside in predominantly Hispanic societies.

6.4. AUSTRALIA POPULATION DIVERSITY

In Australia, surrounding, populace, and maintenance whirl around three associated but majorly separate considerations: the immigration/carrying capacity argument, contrasting betwixt ancestry and Anglo-European approaches of environmental cooperation, and analysis on home maintenance crisis. Each of these filaments visits in its way to contrast and difference, and each has identified that ethnicity plays an essential part in shaping one's perceptions, including their approach towards nature, environmentalism, and maintenance. Yet they all show swank tendencies. Arguments over immigration and ferrying capacity fixate on the analytical moderately than the cultural, often locating migrants as an environmental burden—specifically when they settle in already densely populated urban locales. Chances to explore the differing environmental volumes they carry with them have therefore been sidelined. Meantime, with essential exceptions, arguments of how Australians describe and entangle with this continent's 'natural' countryside and resources have insufficiently looked farther the ancestry/ Anglo-European twofold, to the mixed pieces of knowledge delivered by migrants from the Middle East Pacific, Africa, the Indian subcontinent, and Asia. Therefore, the Commonwealth Government's currently launched maintainable population plan specifically noted the function of ancestry pieces of knowledge in environmental and the preparation strategy, but not those of mixed migrants. And, while the cultural environmental investigation in suburbs of Australia has made essential advancement in switching the focus of environmental arguments from 'out there' to the home scale, it has endured dominantly (although inevitably) concentrated on the Anglo-European ethnic most (Kopka et al., 2009).

This section plans to deliver together these three filaments of arguments and familiarize them in beneficial methods around diversity ethnicity. More than 10 years ago, researchers pulled attention to space in knowledge about differing 'methods of viewing the Australian natural world' brought by mixed migrant associations. But, to take total benefit of the opportunities allowed by ethnic diversity, environmental inquiry and rules require showing far bigger ability and openness to the mixture. Here, we discuss for bigger acceptance of the cultural capital given by mixed Australians, as our community challenges a range of intricate and pressing the environment immigration and ferrying capacity.

Neo-Malthusian agues over the excellent size of the populace for this continent have been a pre-profession of Australian scholars and policymakers since Griffith Taylor. Arguments have concentrated on the environmental maintenance of Australia's general immigration absorption, as well as its dispersion through the continent worry about the size of Australia's populace were refreshed with the forecast in the 2010 intergenerational account that there would be 35 million of us by the year 2049. This forecast arouses a glut of media concentration and public controversy, which saw the classification of a Federal Minister for Populace for the first time in 2010 and the dispensation of Australia's first maintainable population plan in 2011. Immigration has made a considerable improvement to Australia's populace.

Conferring to the 2011 Census of Populace and Housing, 25% of all Australians are earliest in order of generation migrants and analogous percentages of having abroad born parents (Figure 6.3).

Figure 6.3. The cultural mix of the Australian populace has transited considerably over time, and by the year 2011, purely 57.7% of Australian inhabitants claimed only Saxon ancestry or Anglo-Celtic.

Source: https://www.businessinsider.com.au/its-official-australia-is-the-most-expensive-country-in-the-world-for-smokes-and-a-few-beers-2016-5.

The dominant ten origin nations of migrants to Australia in 2010–2011 mirror this rising diversity: China, South Korea, United Kingdom, South Africa, Ireland India, Philippines, Vietnam, New Zealand, Sri Lanka, and Malaysia. Net abroad migration has devoted anywhere betwixt 43.1 and 66.2% of the general yearly populace rise from the year 2000. Necessarily, 80% of freshly migrants settle in Australia's central cities, which previously face a matter of 'livability' problems and populace strains such as urban drape, congestion of traffic, strains on resources of water and affordability of housing. Interpreters of populace rise have portioned much of the accountability for these obstacles on immigration. Populace policy structures have also aimed (debatably imbalanced) on immigration—not only due to its analytical importance but also due to these are the simplest levers to adapt. The rates of fertility and patterns of settlement of the inhabiting populace are far difficult to change. The Governments of Commonwealth maintainable population plan and the Change of Demographic and Livability board report indicated that plan struggled to bring some equilibrium into Australian populace-environment arguments, by accepting that immigration and populace advancement bring both problems and chances, and that populace is not the reason of, or answer to all of Australia's problems. Yet populace-environment arguments continue to bring 'heated debates, emotion, and too often discrimination and racism.' In Australia, as in Northern America, environmental defenders of declined immigration have been blamed for racism—or at the very least, of playing into the hands of associations with honestly racist programs (Kopka et al., 2009).

In the meantime, scholars, and commentators worried over the probability for anti-immigration beliefs to a quirk in the name of environmentalism have been debited with inappropriate racism label in a reason of environmental argument. While this stalemate is undisputedly annoying, it would be ignorant to assume that considerations over the environmental impulses of immigration could overthrow such allegations—given the historical and current racialization of immigration considerations and the Eurocentrism /Anglo of prevailing environmentalism. Western movements of environmental have been broadly blamed for the requirement of cultural diversity in their leadership programs; the Eurocentric/Anglo, intermediate-class environmental worries and knowledge of nature that they prefer; their decline to show other forms of environmental knowledge; and for criticizing environmental degradation on immigration moderately than western behaviors.

The positioning of immigration as an environmental matter in Australia has essential consequences for cultural minority migrants—who's right to attend the national space has been progressively queried on ethnic grounds, and who are now challenged with the additional depression of blame for environmental sufferings in slogans. A rising again of Australian nationalism and current population environment argues are complexically joined. Environmental worries have been all too easily used by those with more outrageous ambitions, as 'a crenelated environment becomes an analogy for crenelated Australian women. Therefore, the anti-immigration attitudes of inveterate racist groupings have been helped by the alternative to environmental considerations. In March year 2010, Australian white dominance association 'Stormfront' (which explains to the motto 'Pride of White, World Wide') used its website to severely encourage the investigative report Dick Smith's Populace Puzzle, which had debated (although in color-blind conditions) for immigration decline on environmental domains. Possibly identify the basic danger of obtaining such revolting supporters; few of Australia's leading environmental associations (containing the Greens of Australian and Friends of the Earth) have been more secretive to name immigration-led populace advancement as the reason of environmental degeneration. There is, therefore, something of a stalemate, which will only aggravate as climate transition provides to the numerous other factors pushing human movement. In such a situation, we hope that ethnic environmental investigation gives something of a circuit-breaker. While not an elixir, such investigation holds positive potential to reveal, and in turn advertise, the exclusive environmental principles, skills, and pieces of knowledge of mixed migrant associations. It gives chances to rattle the 'challenge rank so often given to migrants in environmental discussions, and to underrate the legitimacy of those who use environmental worries to encourage prejudice (Kopka et al., 2009).

6.4.1. Bringing Cultural Diversity into Australian Maintenance Arguments

A short analysis of three major fields of environmental argument shows their cultural particularity and the range to which they have been inboxed by an aim on numbers of immigrants. But we want to view outside limits to ask what other outlooks are attainable. Our catalyst is binary: to mediate effectively in the anti-immigration oratory that has risen in the label of environmentalism, and to promote a manner of environmental reasoning that is ajar to contrast, exigency, and amazing possibilities.

Paul Ehrlich, author of The Population Bomb—a famous neo-Malthusian and boisterous analyst of growth of population at the state and worldwide scale—has understood the essentiality of an ethnic view: What is severely required now is much better acceptance of the ways in which ethnicity changes and regulates human race treatment of its life backing arrangements. We require understanding of how ethnicity evolution gives out the large diversity of human natures—contrast elemental approach, proclivities, beliefs, behaviors and preferences. That should aid us to uncover how to rearrange communal, political, and economic inducement and cut across obstructions of denial and ignorance, permitting the community to turn onto a way toward continuity.

6.5. ASIA POPULATION DIVERSITY

Asia can change its diversity into a pulling force for evolution in a globalized still broken globe, the Laos Prime Minister said on the other day. "Asia has been realizing fast advancement and has remained a locale of great racial mixture," Thongloun Sisoulith told the prospective of Asia convention in

Tokyo. The Prime Minister said that irrespective of the locale's rapid growth, "there are still many nations being classed as least advanced nations" and that "the space between the lower class and the upper class persists to expand."

Divisions betwixt upper class and lower class, as well as cultural brawls between ethnic and religious associations, are scarcely rare to Asia. Thongloun indicated this brings a chance for the locale to prove to the globe a way ahead. "Diversity should not be seen as a barrier. On the antithetical, it should be noted as a chance to augment locational firmness for Asia in order to accommodate the good in the transformation of globalization," he said. "Nations in Asia should change the locales rare idiosyncratic and diversity into a pushing force for advancement of mutual effort." Asia can become a focal point of intelligence, a sharing, and caring society, moving ahead into the age of worldwide integration and assimilation in a favorable and energetic aspect (Prodi, 2002).

The Prime Minister also accentuated the requirement for peace and protection as essential for communal and commercial advancement words that hold more weight when one acknowledgment of how Laos was destroyed by civil battles in the 1950s across the 1970s. Currently, Asia faces accelerating compel from terror attacks, continued harassment by North Korea and boundary pressures betwixt China, and its acquaintance. "Asia must aggravate its exertion in guaranteeing peace and cohesion by resolving all challenges by serene ways, averting all ways and blackmail of armed forces or the use of battalion," Thongloun said. "The most outstanding way to assure resistance to some degree and cohesion in the locale is through the building of confidence and mutual understanding enhancement. Through the serene exchange of ideas and bargaining," he said, adding that nations should awe one another's countries supremacy and boundary uprightness."

An investigation was grounded on scrutiny of men and women working in the extent of companies. The scrutiny was done in March and April this year with colleagues grounded in China Mainland, Japan Hong Kong Malaysia, and Singapore. A sum of 58% of colleagues said they account for a male-line administrator. That is advancement in the past year when 63% of colleagues account for a male-line administrator. Malaysia had the biggest percentage of colleagues account for a female line administrator (46%) and lowest in Japan (28%). Nevertheless, the analysis found there was a view amongst a meaningful percentage of colleagues that approach to jobs, career, and play opportunities for those of equitable capability could be restricted by determinants for instance as disability, age, gender, ethnicity, marital status, family commitments, religion, sexuality or race.

"Earnestly backing for diversity is having a bigger impulse on the aspect of an employer's type amid current employees and hunters of job," announces Simon Lance, managing administrator of Hays Bigger China approach to job opportunities in their company. Hong Kong had the biggest percentage of colleagues who do trust job opportunities are accessible evenly irrespective of the race, ethnicity, or even religion (39%). Singapore had the biggest number of colleagues who trust approach to job opportunity is not uniform in their companies irrespective of the race, ethnicity, or even religion (25%) (Prodi, 2002).

Uniform pay was another spiky matter with only 30% of colleagues throughout all of our performing markets trusting employees of uniform capability are compensated uniformly. Employers got the most excellent evaluation from colleagues in China Mainland where only 12% said compensation and rewards were not awarded uniformly to those of the same capability in the culturally and ethnically diverse locale. Colleagues in Singapore were again the most discontented with 26% announcing uniform allowance for equal capability was not the case in their company. At the same time, 63% of colleagues trust their leaders are genially biased towards workers who think, act, and look, as they do. "The accelerating intricacy of most companies currently means organizations require a consistent pipeline of fresh ideas and ways of doing things if they are to achieve an aggressive benefit. A mixed force of work is the key to assuring an organization can not only adapt at transiting circumstances but influence that transition to continue building on their favorable outcome," says Lance.

Employers require to do more to check comatose bigotry when gathering resources and advancing. Equal compensation also requires effective backing to assure candidates and employees together have the assurance that hard work and powerful accomplishment are the keys to the favorable outcome—not being of the alike gender or culture as their administrator. More than 50% of all colleagues said they had a competent bias at some point in their job. Amid our managing markets, the highest percentage of colleagues who felt their possibility of being acknowledged for a career had been impulse by a factor external capability were in (62%) in Singapore, pursued by (55%), Hong Kong, (50%) Malaysia and China Mainland and Japan (both 49%) (Prodi, 2002).

6.6. ANTARCTICA POPULATION DIVERSITY

Antarctica is the globes extreme south continent taking rest in the Circle of Antarctic and encircled by the South Ocean. Containing 14 million kilometers square of extent, it is the fifth vastest continent. Approximately the mainland's 98% is concealed by the depth of ice 1 mile, which assuredly creates for the severe landscape that has little entrenched populace. In the year 2016, the biggest populace documented in the year is approximately four thousand (Figure 6.4).

Figure 6.4. Antarctica population.

Source:https://www.coolantarctica.com/Antarctica%20fact%20file/science/can_you_live_in_antarctica.php.

The populace of Antarctica is composed chiefly of scientific investigations workers. The inhabitant's number changes, from approximate 1,100 in the severe Antarctic winter to approximate 4,400 in the temperate hot season from October–February, besides, supplementary workers of 1,000 in the adjoining waters. It has no inborn citizens, only long-lasting and summer workers at its many investigation centers. Forward with the 1,100–4,400 investigation workers, there is mostly a complementary 1,000 people, containing ship's crew and scientists operating in-depth analysis in the waters of the agreement Antarctica's locale. The book of facts of the global Central Intelligence Agency keeps a revised disintegration of Antarctica's populace. A total list may be seen across the book of facts of globe by CIA, these are the nations with approximately in Antarctica's people in the crest summer: Chile was 359, Argentina was 667, Australia were 200, France were 125,

Italy was 102, Japan was 125, Russia was 429, the United Kingdom was 217 and United States were 1,293 (Bonner and Walton, 2013).

Even though broadly reasoned of as not being possessed by anyone, Antarctica is demanded by seven differing nations- the Australia United Kingdom, France, New Zealand, Chile, Norway, and Argentina -- and all of these nations cultivate Antarctica a grounds. Brazil, Russia, and the USA have all constrained the right to demand Antarctica domain in and even though they have not already formed an allege, they also control each long-lasting existence. With numerous authorities controlling long-lasting fortified study adroitness here, there is a minute but not well-balanced populace betwixt 1,000–5,000, relying on the period. This indicates Antarctica's populace density to be only 0.00007 people in kilometer square to be 0.0035 people in kilometer square.

6.6.1. Population History of Antarctica

The first genuine semi-long-lasting inhabitants of the locale in Antarctica were seal catchers from America and Britain who used to spend approximately a year in the locale from 1786 progressively. In the vast age of whaling up to 1966, the populace of the locale summed 1,000–2,000 in the hot seasons, and 200 in the cold seasons. Almost early habitations contained Edward Point King, Leith Harbor, Ocean Harbor, and Godthul.

6.6.2. Number of People in Antarctica

Due to this, and the existence of scientists from another nation across the globe, Antarctica has an exact cultured populace. USA scientists constitute the biggest single association, but inhabitants of 19 other nations were also inhabitants on the globes most continent southerly in the year 2008. They reside on 37 dissimilar long-lasting Antarctic grounds, the minute of which is central to only six people and the vastest of which has two hundred and fifty long-lasting inhabitants.

6.6.3. Antarctica's Human Activities

Irrespective of Antarctica's size, bigger than the Mexico and United States integrated, its very continuation as a continent was not entrenched certainly up to the 1820s. It was 1899 before humans first wintered on shores of Antarctic and 1911 before Amundsen arrived at the Pole of South. Humans would not again go on the pole up to 1956. Just in the 1930s, with caravans for instance as those of Ellsworth and Byrd, did an orderly and broadly scientific investigation of the locale begin. Just up to 1957–1958, in the first year of International Geophysical Year, did a science-adaptive, international unified work become a phenomenon. Out of this works risen the Treaty of the Antarctic that has been the aim for global scientific unity ever since. Excluding the taking advantage of whales and seals, the activities of humans have left few long-term points in the treaty locale of Antarctic. Caravan and long-term stations uniquely have impacted local commotions in the past, but the present attention is on washing up past challenges and paying more consideration to environmental worries related to Antarctica's activities (Bonner and Walton, 2013).

In present years, the global political and communal climate has generated human activity in the region of Antarctic to advance speedily, elementary for two objectives: (1) countries that historically were not involved in exploration of Antarctic have looked for description in the Antarctic society and have entrenched scientific grounds so that they may also be engaged; and (2) private inhabitants, in accelerating numbers, have visited the Antarctic as guests to appreciate the continent's immaculate and aesthetically attractive surrounding and the adventure spirit acquiring from the Antarctica's remoteness. Activities of humans that may have environmental results in the Antarctic have been properly registered by few authors, most prominently Bonner and Benninghoff, who itemized the typologies of impacts generated by the activity of humans and proposed ways to appraise them in a scientific base work (Bonner and Walton, 2013).

6.6.4. Physical Surroundings

Antarctica is concealed majorly by ice. The change zone betwixt the continent and the environment seas is the site of rare regimes of climate. Severely tenacious offshore katabatic winds are generated by the plunge of cold air riffraff's from the inside of the continent. The transition in relative times of darkness and daylight are also severe— darkness for 24 hours in the cold transition to the daylight of 24 hours in summer (Figure 6.5).

Figure 6.5. The seas neighboring the continent chill during the winter and melt in the summer. This refrigeration and melting make a change in physical surroundings and add to the numerous nutrient loads in the neighboring waters.

Source: https://discoveringantarctica.org.uk/oceans-atmosphere-landscape/ice-land-and-sea/key-physical-features/.

The severity of the Antarctic surrounding is highly taxing on human aspirations. The chronicles of the early trailblazers have the graphic characterization of the climate of Antarctic. One of the most popular of this trailblazer, Captain Robert F. Scott, paraded with four associates drawing sleds to the South Pole, showing up on 18 January 1912. They died coming from the pole but left itemized explanations of the adversities of the trek. On the arrival, for instance, on 19 January 1913, at a height of 2,960 meters, the margin temperature was 32°C. As they advanced their walk towards the coast in the pungent cold, one entrance states, "cold night, minimum temperature –37°C. Later, on 2 March, Scott's Chronicle states, "it's down to –40°C this morning it enchanted one and one-half hours to put our footgear on, but we got away slightly before eight." Due to the brutality of the climate in the Antarctic, any scientific or other activities in the locale need special facilities and equipment. To decline the necessities for human existence, some analysis activities are now contributing approaching to the use of distant, crumbling facilities for scientific investigations and measurements (Bonner and Walton, 2013).

Norwegian trailblazers had arrived at the South Pole before them: "Great God, this is a terrible locale and awful enough for us to have worked to it with no payment of preference." No doubt, even presently, many Antarctic scientists pursuing activities outdoor that do not go conferring to the arrangement have felt these same disappointments. The brutality of the Antarctic surrounding brings an extra length of complication to everyone who visits or works there. People unfamiliar with these circumstances rapidly learn that extra time and effort are needed to achieve even simple tasks. Even though investigation activities—first preliminary and today more widely concentrated on a sphere of scientific boundaries—has been dominant Antarctica's human activity for approximately half a centurion, this has not consistently been the case. Antarctic waters backed a considerable sealing and whaling industry during the 19th–20th centuries. Even though sealing and whaling are not now done, these waters still back economic fishing for a collection of species. Possible the most famous transition in use of the Antarctic is the notable rise in numbers of visiting tourists the continent in current years.

6.6.5. Research, Resources, and Exploration

Alone from the caravans of the early 1900s, progressing human activity in Antarctica started in the early 1940s, with people of few countries overwintering annually. Beltrami has accumulated data on the numbers of bases and summer and overwintering people across 1990 bases performing in Antarctica advanced from 6–40. Since the early 1800s, humans have used a variety of species that live the seas of Antarctic. Usage of elephant and seals fur started in the early 1800s and progressed up to 1960. Just before 1900, whaling in the Antarctic became a very wide, global industry and, excluding the years of II World War, progressed into the mid-1980s. The population of whale by then had declined to tremendously low numbers; under force from the opinion of the public, the Commission of Whaling International impose a moratorium on the economic take of whales. Specifically, in the region of Antarctic Peninsula, numerous beaches, more so those near to whale conversion facilities or moorings used by vessels of whaling, accommodate big bones of the whale as mute testimony to this last exercise. In current times, Antarctic fish and krill have emerged more essential economically. The seas of Antarctic are far from uncontacted. Their biological assets have been harvested broadly, and few species have been considerably depleted. This harvesting is probable to progress, specifically if krill emerges an essential point of supply of protein for humans and/or other consumptions. The consumption of fish and krill is now controlled under the conference for the preservation of the Living Resources of Antarctic Marine (Bonner and Walton, 2013).

6.7. AFRICA'S POPULATION DIVERSITY

Africa is the next-biggest and second-best populated continent on the globe with an approximated populace with 1.2 billion people in the year 2016. Africa being home to fifty-four acknowledged autonomous nations and nations, 9 regions and two liberated nations with very small identification. The United Nations (UN) Population Fund indicated in the year 2009 that the populace of Africa had hit the strike billion point and had henceforth increased in the amount above the duration of twenty-seven years (Figure 6.6).

Figure 6.6. Africa population diversity.

Source: https://www.slideshare.net/above69/yahclick-africacom-1719-november-2015-cape-town-south-africa.

The Director of Fund's Populace Thoraya Obeid said to the BBC at the duration and emphasized the fundamental objectives of the advancing populace. "Nations of Africa are all rising rapidly due to there is a big number of females who have no entry to families preparation," she said. "It is an African anomaly of a big advancing populace and a big proportion of young people in the populace."

6.7.1. Rise of Populace and Anticipation of Life in Africa

While the growth of population is almost low in some locales, nations, for instance, Uganda and Nigeria are advancing at a developed rate. In many nations in the continent, the rise of the population is in surplus of 2% annually. In extension, there are a better percentage of young people in the African populace as a total, which indicates that 41% of the populace in Africa is below 15 years of age. The expectancy of life is also little - less than fifty in many countries and equitably 52 throughout the continent as a total. This has declined majorly over the past 2 decades with extensive AIDS and HIV endemic taking plenty of the criticism for that number. Toddler's deaths are also severely great, and in the nation of Mali, sadly, there are 102 mortalities in a thousand live deliveries. All of these figures could likely lead to a decline in the numbers of the populace (Payne et al., 2016).

6.7.2. Demographics of Africa

The countries of African as totally are made up of such a differing set of constituents that it is absurd to list them in full involving demographics. Nevertheless, in various locales of the continent, there has been a rise in European and Asian colonizers, which has also dressed to increase the populace enumeration as a total. In earlier colonies of British, this can be viewed progressively. Uganda, Tanzania, South Africa, and Kenya are all good illustrations as to a rising set of different tribes. The African populace has accelerated fast over the past 40 years and it has a comparatively young populace, with more than 50% of the populace below 25 in some nations.

6.7.3. Advancement of Africa Populace

Any professional would find it serious to contend with the generally held perception that the African populace in the year 2016 and afterward is set for farther acceleration. With small or no laws in locale to address the matter, the projection of the 2.4 billion by 2050 is completely feasible.

Africa now has a much small density of the populace of approximately 65 people in mile square, which puts it below Europe, South Africa, and Asia. The African populace is now estimated to quadruple in just ninety years, with a ratio of growth that will make Africa more essential than ever to the worldwide economy.

Nigeria is now one of the most populated nations on the globe, and as China's population decreases and India plains, Nigeria will reach approximately 1 billion people by the year 2100 and come near to exceed China. This is attractively amazing looking the nation almost at the size of Texas. Nigeria is set for one of the vastest populace explosions in global history and it is projected to rise by a factor of 8 in just 2 or 3 generations.

The explosion in the population of Africa will be in sub-Saharan, counting progression in nations like Tanzania, which is one of the destitute nations on earth. Just thirteen years ago, the populace of the nation was 34 million, which has now risen to 45 million but is expected to range 276 million by 2100, which is near to the current populace of the United States (Payne et al., 2016).

Many see the population in Africa's rise a bit fearsome, with forecasts placing the continent's populace at 2.4 billion by the year 2050. By the year 2100, more than half of the global growth is projected to come from Africa, embracing 4.1 billion people by 2100 to demand below a third of the populace of the world. Most nations will at least triple in the populace as the locale has very high rates of fertility and very small family planning in major locales. As much of Africa is still advancing, and it includes some of the destitute nations on the globe, time will tell how it will maintain such large growth of population.

6.7.4. Africa's Human Evolutionary History

Africa is a locale of abundant cultural, linguistic, genetic, and phenotypic differences. There are more than 2000 different ethno-semantic associations in Africa, talking accents that comprise almost 33% of the languages of the world. These populaces exercise a broad range of livelihood designs including various types of gathering-hunting, pastoralism, and agriculture. Africans also live in atmospheres that sphere from the largest globe desert and next vastest equatorial rainforest to marshy places, highlands of mountain, and savanna, and these atmospheres have, in few cases, underwent exciting transitions in the current history (Payne et al., 2016).

Conferring to the Africa design of contemporary origins of human, anatomically, contemporary human-beings introduced in Africa and then advance through the rest of the world within the past approximately 100,000 years. The change to contemporary humans in Africa was not abrupt; rather, the paleobiological file shows an erratic mosaic of contemporary, regional structural, and archaic and observable traits that happened over an important duration of chronology and through a wide geographic sphere in Africa. The first famous obtained suite of morphological attributes related to contemporary humans arrives in fossil remnants from Ethiopia. Nevertheless, this analysis does not rule out the reality of contemporary structural attributes in other locales of Africa, specifically where samples may be less well conserved and/or where broad paleontology and paleobiological findings have not been administered. A multi-locational ancestry ideal for contemporary humans in Africa is not as probable as it would be for the populace globally, looking at the bigger promising for movement and blending in a continental single locale. A more advanced contemporary suite of attributes seems in Asia Southwest and Eastern of Africa, pursued by a fast expansion of contemporary humans across the rest of the African continent.

Two migration courses of contemporary humans in Africa have been projected. The existence of contemporary humans in Africa as early as approximately 50 Kya, which precedes their existence in Africa approximately 40 Kya, has indicated a south coast way in the Indian Ocean in which contemporary humans first entered Africa (through Ethiopia probably) by overpassing the Bab-el-Mandeb difficulty at the opening of the Red Sea and then fast moved to other parts of Africa. This movement is backed by the existence of many aged haplotypes in Southern Africa and their inexistence in the Levant. Other imitators have culturally popular the next northern way through the Peninsula of Sinai to the Levant. Irrespective of the way of movement of contemporary humans of Africa, the shared designs of genetic diversity across non-African populaces and the different patterns of genetic abnormality among the population of African talk against repetitive examining of the diversity of African from multiple populations source. Nevertheless, studies of more autonomous position and a vast number of the populace of African, specifically from Eastern Africa, will be required to better project the number and ancestry of migration affairs in Africa. After contemporary humans moved to Africa, probably could have been some blending of contemporary humans with the ancient populace (Payne et al., 2016).

The movement of contemporary humans to Africa is reasoned to be escorted by a populace barrier. These projections indicate that Africans must have fast extended to a bigger size to detail for projections of a long-term active size of the populace of approximately 10000 individuals for populations globally. Few current investigations show a fast advancement of African populations in the last approximately 50,000 years.

6.7.5. Genetic Diversity Patterns in Africa

The design of genetic differences in contemporary populations of Africans is affected by a history of demography (for instance transitions in size of the population, short and long sphere movement affairs and blending) as well as locus-definite forces such as choosing naturally, and metamorphosis recombination. For instance, the movement of Bantu accents agricultural from Western Africa across the African sub-Sahara in the past approximately 4000 years and resultant blending with ancestry populace has had a dominant impulse on the arrangement of change in the contemporary populace in Africa. Though Africa is demanding for knowing contemporary human ancestries and genetic danger factors for infections, it has been under-symbolized in human genetic investigations. Much of what we now know about genetic differences are from a narrowed number of the approximately 2000 ethnic-linguistic associations in Africa, and the bulk of this information is from Y chromosome and mitochondrial DNA investigations. A bigger-scale autosomal investigation on the diversity of African genetic is only now starting to become accessible (Payne et al., 2016).

7

URBAN, URBANIZATION, AND HUMAN POPULATION

CONTENTS

7.1. INTRODUCTION

The description of 'urban' changes from nation to nation, and, with occasional re-categorization, can also change inside one nation over the duration, making precise contrastingly hard. An urban locality can be described by one of the following: administrative principle or civil barriers (for instance locality in the administration of a municipality or town task force), an opening size of populace (where the margin for an urban settlement is originally in the locality of 2,000 people, even though this varies worldwide betwixt 200–50,000), density of population function of economy (for instance where a meaningful majority of the populace is not elementarily involved in agriculture, or where there is excess employment) or the presence of urban aspects (for instance, concrete streets, sewerage, lighting of electricity). In the year 2010, about 3.5 billion people resided in localities categorized as urban.

An urban context can be described widely on the ground of density of population, consolidation of bureaucratic bodies and infrastructure and a different set of livelihood and money creation activities. Urban localities will be noted by the high density of population when contrasted to other localities. While some cities are described by municipal barriers, the majority of urban centers have not been entitled as such. They are essentially distinguished by the presence of bureaucratic structures, for instance, government offices and courts and a relative consolidation of services for instance as commercial institutions such as banks and hospitals. In an urban context, the patterns of occupation and income creation activities will be different and different in rural localities not majorly bound to the production of agriculture. If the locality in inquiry fits some if not all of these elementary components, it can be observed as urban (Birch and Wachter, 2011).

Urbanization (or urbanization) attributes to the shift of population from rural localities to urban localities, the continuous rise in the percentage of people residing in urban locales, and how each community acclimates to this transition. It is dominantly the transformation by which cities and towns are created and become bigger as more people start residing and functioning in central localities. Even though the two ideas are sometimes used correspondently, urbanization should be differentiated from the growth of urban centers: urbanization is the percentage of the total state populace residing in localities categorized as urban while urban growth refers to the aggregate number of people residing in localities categorized as urban. The United Nations (UN) calculated that 50% of the global populace would reside in urban locales at the end of the year 2008. It is projected that by the year 2050 about 64% of the advancing nations 86% of the advanced global nations will be urbanized. That is proportionate to about 3 billion people residing in urban by the year 2050, much of which will happen in Asia and Africa. Prominently, the UN has also currently calculated that almost all world growth of population from the year 2017–2030 will be by cities, approximately 1.1 billion fresh people residing in urban for the next decade.

Urbanization is important to a variety of subjects, including the planning of urban locales, architecture, sociology, geography economics, and the health of the public. The anomaly has been closely related to innovation, growth of industries, and the sociological development of rationale. Urbanization can be seen as a definite circumstance at a set time (for instance. the percentage of total populace or locales in towns or cities) or as a rise in that circumstance over time. So urbanization can be evaluated either in terms of, say, the urban advancement level analogous to the general population or as the amount at which the urban percentage of the population is accelerated. Urbanization makes many communal, commercial, and environmental transitions, which create a chance for sustainability with the "power to use resources more favorably, to make more tenable land use, and to guard the biodiversity of the natural environment."

Urbanization is not purely a recent anomaly, but a fast and historic metamorphosis of human communal roots on a world scale, whereby dominantly culture of rural areas is being fast reinstated by the broadened culture of urban. The first major transition in settlement arrangement was the aggregation of and hunters into villages many thousand years ago. Village perception is noted by general bloodlines, confidante relationships, and social attitudes since urban perception are signified by obscure bloodlines, bizarre relations, and aggressive behavior. This extraordinary migration of people is out looked to advance and aggravate during the next few mercenaries, upcoming cities to capacities absurd only a century ago. As an outcome, the global urban growth of population trajectory has up till currently pursued a quadratic-hyperbolic figure.

Today, in Asia the urban collections of Mumbai Karachi Osaka, Jakarta, Beijing, Seoul, Manila, and Shanghai are each previously home to over 20 million people, while Delhi is projected to access or outstrip 40 million people by the year 2035. Cities, for instance, Istanbul, Tehran, São Paulo, Mexico City, New York City, London, Cairo, and Lagos are or shortly will be, home to more than 10 million people each (Birch and Wachter, 2011).

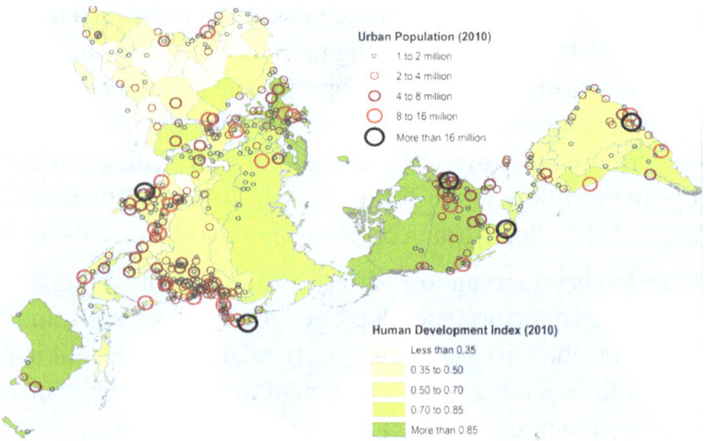

Figure 7.1. World's Largest Cities and Human Development Index 2010.

Source: https://www.researchgate.net/figure/Worlds-Largest-Cities-and-Human-Development-Index-2010Source-UNEP-2011_fig3_263557449.

Urbanization is not simply acceleration in the percentage of urban populace, but it is an ongoing course that entails all aspects controlling the process of growth of the economy and communal-cultural transitions. The rise, spatial escalation, advancement, and decrease of cities and towns are very much intertwined with the methods of urbanization. These methods are triggered because of differing enforcing factors in different moments. Thus, one can't evaluate the current with any appropriate knowledge of the factors accountable for the advancement and rise of urban centers in the past. An effort has been made to investigate the experiences of urbanization in Murshidabad district from prehistoric duration to current time along with the consideration of the major processes involved in the historical durations. It also aspires to mark out the spatial and intern stops in urbanization which has been an essential attitude of urbanization from past to current times.

7.2. HISTORY OF URBANIZATION

Urbanization is the mechanism by which rural societies grow to form cities, or urban centers, and, by continuation, the progression, and development of those cities. Urbanization started in old Mesopotamia in the Uruk duration (4300–3100 BC) for objectives scholars have not yet come to unison on. It is contemplated, nevertheless, that a specific flourishing and capable village drew the consideration of other, less flourishing, tribes who then linked themselves to the successful settlement.

Lewis Mumford, the historian states that: "though long-lasting villages are dated only from times of Neolithic, the custom of being used to caves for the general accomplishment of magical celebrations seems to backdate to a previous duration. The city's outline as both an external form and an outward arrangement of life might be found in such convergence."

This transformation, then, gave rise to the centers that are populated densely which came to be acknowledged as cities. Chapin Metz Hellen, the historian suggests that the progression of the cities in Mesopotamia was an outcome of the residents straining to contend with the environment. She writes: "The civilized life that appeared at Sumer was shaped by 2 constraining aspects: the uncertainty of the Euphrates and Tigris rivers, which at any time could discharge degrading floods that swept out almost all peoples, and the severe productivity of the valleys of the river, led by centuries-old deposits of soil. Thus, while the river valleys of Mesopotamian south drawn the movement of neighboring peoples and made easy, for the first time in the past, the rising of excess food, the evaporation of the rivers required a form of composite management to secure the soggy, low-lying land from flooding. As excess production accelerated and as composite management became more progressive, a process of urbanization derived and the civilization of Sumerian took root."

7.2.1. The Rise of the City

The earliest city to progression in the location of Mesopotamia is seen by current scholars to be Uruk, approximately 4500 BC, and Ur around 3800 BC was next, both of which were then located in near to the banks of the River Euphrates. To the Sumerians, nevertheless, the earliest city was Eridu which was established in 5400 BC but likely was not a city in the alike way that Ur or Uruk would be explained. In the Inanna myth and the God of Wisdom, the states of culture noted as the meh, are obtained from Eridu to Uruk and it is perceived the discipline of this poem shows the change in the example from a more agrarian way of life to one more citified. The arrangement of the city and the protection of urban residence seem to have drawn the population of the locale to city centers although the approach has been indicated that the populace was energetically detached from holdings of agriculture and re-situated in the cities whose rulers then suitably the lands for the nation. This approach, nevertheless, fails to account for the progression of urbanization across the past of Mesopotamia or its reproduction in other countries (Birch and Wachter, 2011).

By the year 2600 BC, Ur was a burgeoning capital and, by the year 2900 BC, was a fortified city with a populace of about 65,000. Urbanization, nevertheless, rose as the city extended out from the center and, in time, the once productive lands which fed the population were infertile. The land over-use, together with a concealed change in the Euphrates which pulled the waters away from the city, the outcome in the complex lastly being left approximately 500 BC. Eridu, for possibly similar objectives, was left in year 600 BC and Uruk in 650 BC. No doubt, many reasons are provided to the decrease of cities such as Ur (Akkad from Sargon ravaged the city in 2340 BC), it has been indicated that urbanization specifically, the exhaustion of the surrounding lands for agriculture, was a major caused. (Thomas et al., 2013) (Figure 7.2).

Figure 7.2. The rise of ancient Egypt.

Source: http://www.illustratedcuriosity.com/history/history-ancient-history/video-rise-of-the-city-the-lost-cities-of-ancient-egypt/.

7.2.2. The Overreach of Resources and Urbanization

At the center of Ur, as with all of the cities in old Mesopotamia, was the enormous temple which was the location of commercial, communal functions, and ceremonial. Religious exercises, such as festivals, were the major communal assemblies of the time and these changes were often used to circulate excess food and excess to the population of the city. The temple priests, who were also the city rulers from about 3400 BC, benefited for this circulation and relied majorly on the farmers or the locale to give such excess as they required (a part which would finally be taken over by the king, as accepted royalty and the religious class is in power during the 3rd millennium BC with the rise of the warrior-king noted as the Lugal). This surplus production nationwide did not only gave the populace of the city with food but also accelerated long-distance businesses with other urban along the Euphrates, for instance, Tikrit and Eridu. As urbanization progressed, nevertheless, the requirement for more and already rawer materials diminished the natural resources of the locale and, finally, led to inadequate necessary assets and the leaving of the city (Benna et al., 2016).

7.2.3. The Response of Egypt to Urbanization

Urbanization escalated from Mesopotamia to Egypt and, from Egypt to Greece and it implies, early on, that the teaching of the city of Ur, and others, was observed by later urban centers. In Egypt, chiefly, great caution was taken with the land to avoid the less attractive results of urbanization from collapsing the great cities of Pharaoh so that point could endure constant on cultural factors, for instance, the writing development, laws, architecture, sanitation, administration craftsmanship, and trade (all perceived to have ancestry in Uruk, Mesopotamia). University of Washington, Professor George Modelski, notes:

> "Some students of the old era have been noted to debate that, dissimilar Egypt, Mesopotamia, didn't have anything that could be viewed as cities in current terms. That vast nation did have cemeteries temples, and palaces, often of monumental percentages, as early as the fourth and third millennia, but its capitals seem to have required unusual size and have left insufficient proof either of academic life or economic activity. As Wilson John puts it: 'For almost 300,000 years, until the basing of Alexandria, old Egypt was a civilization outside a single namable city."

This assertion, nevertheless, is opposed by Arizona State University Professor M.E. Smith who says that:

> "Because archaeologists have declined to find vast cities in Egypt before the capital of Akhenaten's at Amarna in the New Kingdom period in year 1350 BC Egypt has sometimes been opposed to Mesopotamia as a `development without cities.' This label masks a different form of urbanism, nevertheless...Egypt did not need cities; rather its urban arrangement was arranged variously from the more known form of Mesopotamian cities."

It assumed that both the advantage and the expense of urbanization and chose for "scattered urbanization described by smaller, more specific urban settlements." This same example holds accurate for the Maya urban centers, at least in their designing, but the implications general advancement of urbanization led to the diminishing of natural resources (Benna et al., 2016).

7.2.4. The Rise and Fall of Cities

The unreal environment of the city, which conquered the neighboring natural environment to the requirements of the population, typically is seen to finally diminish and damage the very resources which gave increase to the city. As urbanization advanced, rural lands declined and, as Mumford notes:

> "The unseen forces of urbanization, gushing along the lines of slightest opposition, show no predilection for making an urban and industrial arrangement that will be constant, self-maintaining, and self-regenerating. On the opposite, as overcrowding condenses and separation widens, both the urban and the rural countryside go through demolishment and deterioration, while the loss of investments in the amendments…serve only to advance more of the taint and disorder they look to cover up."

This phase of rising and fall of cities is seen repetitively in many nations around the globe. Why it occurred so commonly in some locales, for instance, a Mesopotamia, and not in others, such as Greece, is a quiz still argued by historians and scholars. Some affirm it is an element of crowding while others say to a deterioration of the natural resources. Neither response is sufficient and most probably it is a composition of many factors, an inadequate forethought among them, which led to the annihilation or forsaking of so many old cities.

7.3. CAUSES OF URBANIZATION

Communal factors, industrialization, innovation, and modernization, and commercial opportunity lead to urbanization or acceleration in the number urbanization are issued in subsections.

7.3.1. Migration

People move from rural locales to the town, industrialized locales due to the comparably better employment chances accessible there. As localities become more productive and affluent because of the mineral's discovery, exploitation of resources, or activities in agriculture, cities begin to develop as the rural locales change to urbanism. The acceleration in output leads to commercial advancement and better value-added job chances. This bears about the requirement to flourish excellent infrastructure, better academic institutions, better facilities of health, and better networks of transport, installation of institutions of banking, excellent governance, and good housing. As this takes place, rural societies begin to embrace the urban culture and finally become urban centers that progressively to advance as more people move to such locales in search of a good life. Many localities have become more productive and prosperous because of the minerals discovery, exploitation of resources, or certain activities in agriculture, therefore, cities begin emerging. It is a well-known truth that the rise in output leads to commercial advancement and higher value-added job opportunities. This tendency normally contributes to the advancement of land for use in economic properties, communal-economic backing institutions, transportation, and living buildings (Mangalam, 2015) (Figure 7.3).

Figure 7.3. Human migration in the cities.

Source: https://globalpressjournal.com/topics/human-migration/page/2/.

7.3.2. Commercialization

Commercialization and business come with the common view that the towns and cities give better business-related opportunities and returns contrasted to the rural locales. Commerce and enterprise play a major part in urbanization. The allocation of goods and services and economic understanding in the modernistic era has advanced modern retailing institutions and methods of exchange that have enormously given acceleration to the progression of towns and cities. All typologies of trades provide over to urbanization. In the current era, the allocation of goods, services, and economic transactions have advanced many current marketing institutions and methods of exchange out-coming in the fast progression of towns and cities. It is a common view in people's minds that commercialization and economy in towns and cities give better chances as well as returns as contrasted to rural locales.

7.3.3. Industrial Growth

Industrial growth is a considerable cause of urbanization. It has promoted the employment chances. Rural people have moved to cities on the explanation of better employment chances. When the urban populace rate of growth is 40% in India, the rate of industrial growth is about 60% per year. Industrialization is a tendency showing a change from the old economics in agriculture to the novel non-agricultural economy, which makes a modernized community. Through revolution in industries, more people have been captivated to move from rural to urban locales on the explanation of improved opportunities in employment (Figure 7.4).

Figure 7.4. The industrialization has accelerated opportunities for employment by issuing people the probability to work in current sectors in job classes that help to stir developments in the economy.

Source: https://www.livemint.com/Money/ZDkgNlAL5Dla4EkeUpaHFN/Capital-goods-sector-2014-to-be-the-same.html.

7.3.4. Social Factors

Many communal factors such as the allure of cities, better living standard, better academic facilities, and requirement for rank also activate people to move to cities. There are many communal gains given to live in cities and towns. Instances contain better facilities in education, improved residing principles, better cleanliness and accommodation, better care of health, better facilities in recreation, and better communal life in generic. On this explanation, more, and more people are incited to move into cities and towns to get the broad collection of communal gains and services which are not available in the rural locales. There are many social advances given to live in cities and towns. People get good facilities in an urban locale like sanitation, educational, accommodation, recreation health care, and residing standards to lead a better communal life in generic. Because of these objectives, more, and more people are incited to move into cities and towns to get a wide collection of communal aids and services which are not generally present in rural locales (OECD, 2010).

7.3.5. Employment Opportunities

In rural setups, people have to rely on majorly on agriculture for their income. In starvation situations or natural disasters, rural people have to move to cities. In cities and towns, many employment opportunities frequently pull people from rural locales to look better incomes. Henceforth, most people mostly move into urban locales to access well-salaried employments as urban locales have numberless job opportunities in all broadening sectors for instance health of the public, transport, education, recreation, and sports, companies, and business undertaking. In cities and towns, many job opportunities gradually draw people from rural locales to seek better income opportunities. Many job opportunities are there in urban locales that truly pull people residing in rural locales to seek better incomes. Due to this, objective people repeatedly move into urban locales in the look of well-paying jobs. Better value-added employment is being produced by such services as well as companies leading to more job opportunities in urban locales.

7.3.6. Modernization

Urban locales are created by intricate technology, good infrastructure, medical facilities, communication, etc. In urban locales, people also accept transitions in the modes of residence like habits in residential, dressing, attitudes, beliefs, and food. Modernization contributes a very key part in the transformation to urbanization. As urban locales become more technology shrewd together with overly intricate communication, medical facilities, dressing code, liberalization, enlightenment infrastructure, and other communal amenities availability, people believe they can lead a happy life in urban. In urban locales, people also accept changes in the manners of living like residential customs, dressing attitudes, beliefs, and food. As an outcome, people move to cities and the cities rise by gripping the progressing number of people day after day. Modernization and fast transition in the current lifestyle of people plays a very essential part in the course of urbanization (Drakakis-Smith, 2011).

7.3.7. Deficient Town Designs

A more startling factor in the common degradation in the standard of civil services is the advancing sense of convalescence of our designers and administrators.

7.4. EXAMPLES OF URBANIZATION

Urbanization rates change throughout the world. The United Kingdom and the United States have a far greater urbanization level than Swaziland, Nigeria, China, or India. Their yearly urbanization rates are far lazy, nevertheless, since a much smaller percentage of the population is still residing in a rural locale and in the method of moving to urban. Localities that have been distressed by urbanization in these nations in more current duration include:

- Urbanization in the United States has STIRRED the Rocky mountains in locales for instance as Wyoming, Jackson Hole; Colorado, Telluride; Mexico, Taos; Colorado, Douglas County, Aspen, and Colorado. The lake district of northern Minnesota has also been stirred as has Vermont, Florida coast, Jefferson County-the Birmingham, Alabama locality, and the North Carolina Barrier Islands.
- In the United Kingdom, two dominant examples of current urbanization can be seen in Wiltshire, Milton Keynes, and, Buckinghamshire, Swindon; These 2 towns indicate some of the fastest rates of growth in Europe.

7.4.1. South Korea-Seoul

Scanty cities have seen such a fast growth of population as Seoul in South Korea. Beginning at a populace of 900,000 in 1945, the population has risen to above 10 million by 1990. This urbanization explosion brought accelerated revenue and commercial affluence for the city, but it also made new species of challenges. Cremation plants and refuse dumps were made outside advisory from local inhabitants, leading to furious inhabitants and their movement from the locale. Transportation arrangements have not been simple to plan; as challenging transport arrangements have, diverse bus transmits and time tables. Creation also has played a part, as physically progressing city needs heavy architecture, which makes traffic overcrowding. The Seoul government has found it important to work intently with local jurisdiction and nationals to administer these matters.

7.4.2. Continent of Africa

At the swing of the nineteenth centenary, Africa southern Sahara had an overall urban populace of less than 5%, most choosing for more cultural jobs agriculturally. By the year 2000, the number of urbanites reached approximately 38%, with a projected rise to above 45% by the year 2015. The progress of urbanization in Africa is lazy, but it is stable.

Forecasts concerning urbanization of Africa have been erroneous, nevertheless, and this is partly because of to the AIDS endemic, abrupt government overthrows and combats betwixt countries. Times of rivalry have seen a tenacious rural-urban migration flux. However, in 1963 the Nigerian city of Lagos, 665,000 inhabitants rose to almost 9 million inhabitants in 2000 and is projected to attain 20 million inhabitants by 2025, making it the eleventh biggest city in the globe. Urbanization is occurring in Africa, just gradual than formerly projected.

7.4.3. Bloemfontein

She is the main city of the Free State Province. The Free State Province has a populace of 2 958 800. Approving Johnstone et al., Bloemfontein has a populace of 325,000. The basic accents of this province are IsiXhosa 9, 1%. Sesotho 64, 4%; and Afrikaans 11, 9%. She is classed under Mangaung local metropolis. She has a well-entrenched educational, administrative infrastructure and institutional, and the Supreme Court of Appeal houses. She is attributed to as the "City of Roses," the home to some of the province's major triennial educational organizations, not excluding the Free State University and the Central University of Technology. There is a gradual incursion of incompetent people into Bloemfontein from rural locales. The joblessness rate is 40%.

7.4.4. Kimberley

Kimberly is the main city of the Northern Cape Province. It has a populace of over 1 million. The major accents of this province are: Setswana and Afrikaans. Other spoken accents are English and IsiXhosa. The Northern Cape is minerals rich. It falls under the municipality of Sol Plaatje. Kimberly has a wealthy and colorful past. Kimberly is known for her diamonds which were found in the year 1870 in the obstructions of a farmhouse and on bordering farmland. The disclosure of diamonds flashed a diamond scrabble that saw tens of thousands of miners from all over the globe speed to the locale to stake their interest in Colesburg Kopje.

The country's major diamond spots are established in the Kimberley district. In year 1888, the diamond trade was properly well-established with the making of De Beers combined Mines. Tellurian diamonds are also broadened from the shores and the sea betwixt Bay of Alexander and Nolloth Port.

7.4.5. Town of Cape

Cape Town is the main city of the Western Cape Province. It has a populace of approximately 5 million. The basic accents of this province are: Isixhosa, English, and Afrikaans. Cape Town falls under the City Municipality of Cape Town. It is the "Mother City" of South Africa. She is still the commercial hub of the West Cape Province being composed industrial locales such as Montagu Gardens, Epping, Retreat, and Parow. On the western coast, the Saldanha Steel Project has led to accelerated commercial activity. Most of the nation's companies of petroleum and the biggest portion of the printing and publishing companies are situated in Cape Town. Strangers to the Western Cape can unembark at George Airport, Cape Town International Airport, or all the locales of Cape Town, Bay of Saldanha or Mossel.

The towns surrounding are progressing in detection to Cape Town. The city's unique environmental locales, containing her beaches, Peninsula National Park in the Cape Town and Island of Roben are essential tourist attractions that bring commercial activity and encourage job making.

7.4.6. East London

The populace of the province is approximately 7 million. The major accents of Eastern Cape are Afrikaans, English, and Isixhosa. Indians Africans, Asians Whites and coloreds are also seen in this province. East London is under the Municipality of Buffalo City. The sections in East London with the biggest progression capability are commerce, accommodations, and manufacturing, communication, and transport and business and financial services. The urbane businesses of East London and Port Elizabeth are grounded elementary on construction and manufacturing. The province is the center of South Africa's automotive corporations. Few of the global vastest motor manufactures, for instance, Daimler Chrysler, and Volkswagen, For General Motors, have industries in the Eastern Cape.

7.4.7. Port of Elizabeth

The populace and accents of the Eastern Cape Province are the populace of the province is approximately 7 million. The major accents of Port Elizabeth are Afrikaans, English, and Isixhosa. Indians Africans, Asians Whites, and coloreds are also seen at this province municipality. In the year 1993, Port Elizabeth grew into the first urban in South Africa with an autonomous and integrated city council. It is the biggest city in the railroad and the fifth-biggest in South Africa. It has well-advanced amenities: transport of public is enough and it is feasible to maneuver across it by city taxi, on foot, or bus shuttle. The city is an essential center for automobile companies and is a considerable transport center. Its advanced modernistic harbor and airport joining Southern Africa with other countries and global terminals. The graceful projectile Bay of Algoa gives a perfect ambiance for the port of Elizabeth. East London offers evenly friendly harbor facilities. The province is ameliorated by three airports located in East London Mthatha and Port Elizabeth.

7.5. FACTORS AFFECTING URBANIZATION

In advancing nations, urbanization mostly happens when people migrate from villages to reside in urban in the hope of accessing a better living standard. Migration is catalyzed by economic advancement and progression and by technological transition and probably also by rivalry and social interruptions. It is projected by pull factors that pull people to urban locales and push factors that drive people elsewhere from the nation.

Employment chances in urban areas are one of the major draw factors. The majority of industries are situated in urban and give possibilities of high urban incomes. There are also more academic institutions giving courses and training in a broad range of disciplines and skills. People are pulled to a city lifestyle and the 'glistening lights' life of the city. All of these aspects result in both interim and long-lasting migration to city locales (Benna et al., 2016).

Destitute living conditions and inadequate opportunities for salaried employment in rural locales are push factors. Forks are migrating from rural locales due to destitute health care and narrowed academic and commercial chances as well as environmental transitions, floods, droughts inadequacy of sufficiently fertile land, and other problems on rural livelihoods.

Rural to urban movement can be a discriminating process, as some typologies of people are more probable to move than others. One of the aspects contained is gender, due to employment chances vary widely with differing jobs for men and women. Another aspect is age. Young people are more probable to migrate to towns, with more old people and children left in rural locales. Discrimination in migration alters the populace in both the rural and the urban locales. If more men migrate to towns and cities than women, this departs a prevalent female community in rural locales (Benna et al., 2016).

Industrialization, communal factors, innovation, and commercial opportunity lead to urbanization or acceleration in the number of people migrating from rural to urban locales. Mental analysis of urbanization is inured below:

- **Migration:** People move from rural locales to the town, industrial locales due to the approximately better employment chances convenient there;
- **Commercialization:** And business comes with the common approach that the towns and cities give better business chances and returns correlated to the rural locales;
- **Industrial Growth:** Rural people have moved to cities on a reason for better employment chances. When the city population advancement rate is 40% in India, the growth of the industry rate is about 60% per year;

- **Social Factor:** Many communal factors such as the appeal of cities, better living standards, better academic facilities, need for rank also convince people to move to cities. There are many communal benefits associated with living in cities and towns;
- **Opportunities in Employment:** In rural setup, people have to rely majorly on agriculture for their income. In famine situations or natural adversities, rural people have to move to cities. In cities and towns, many job chances constantly pull people from rural locales to look better income. Consequently, the bigger percentage of people intermittently move into urban locales to get well-salaried jobs as urban locales to have numberless employment chances in all expansion sectors for instance education, the health of public, sports recreation, and, transport industries, and business undertaking;
- **Modernization:** Urban locales are described by cultured technology good infrastructure, communication, facilities of medics, etc. In urban locales, people also support transitions in the approaches of living namely livelihood, dressing attitudes, habits, beliefs, and food; and
- **Flawed Town Planning:** A more dangerous aspect in the common degradation in the standard of community services is the advancing sense of convalescence of our designers and administrators.

7.6. IMPACTS OF GROWTH IN URBAN CENTERS

Booming towns and cities are an important aspect of the thriving economy of a nation. The gathering of human and economic assets in one locale encourages innovation and advancement in science, business, industry, and technology. Accessibility to health education and cultural activities is more quickly accessible to people in cities than in rural. In urban, rates of child survival are higher than in rural locales due to better health care access. The urban population's density makes it simple and less expensive for the government and utilities to administer important services and goods (Drakakis-Smith, 2011).

Universities colleges and schools are entrenched in cities to develop human assets. A collection of educational programs are accessible, giving students a broad alternative for their future jobs. People of many categories and religions reside and work composed in cities, which makes good compassionate and harmony and aids break down communal and cultural barricades. Cities also have developed communication and transport systems.

Nevertheless, these many advantages of urban life don't spread to all. Fast increases population and accidental advancement make an urban drape with negative commercial, communal, and environmental results. The urban growth rate often burdens the volume of the local and national governments to support urban livers with even the most elemental services of habitation, a supply of water, drainage, and disposal of solid waste.

7.6.1. Habitation

In advancing nations, approximate 33% of urban inhabitants reside in needy slums and settler settlements. Slums are urban localities that are densely populated and have mediocre housing with very destitute residing circumstances, making several challenges.

In, a report in 2008 Addis Ababa, shown that 80% of the buildings in the city were categorized as slums because of the physical degradation of its habitation, congestion, poor approach, and inadequate infrastructure services (Drakakis-Smith, 2011) (Figure 7.5).

Figure 7.5. Addis Ababa Urban slum.

Source: https://www.compassion.com/ethiopia/addis-ababa.htm.

Slum localities typologically suffer from:

- Destitute housing with small, congested houses built very near together using deficient materials and with the ambivalent supply of electricity;

- Constricted approach to supplies of water;
- Poor Sewage Facilities, which leads to a corrupt and disgraced local environs;
- Insufficient health care facilities which, aggravated with the destitute residing circumstances, rise in sickness and rates of death; and
- Unsafe living circumstances—slum livers may be enforced removed by landlords or other jurisdictions.

7.6.2. Supply of Water and Cleanliness

The supplies of water and lavatory services to developing urban settlements, suburbs, and slum localities present demanding obstacles. The risen water demand from the developing populace can place joined emphasis on already elongated resources. In and about cities, water is generally in short supply and conditional to increasing challenges by differing users. The growth of Urban leads to expanding water demand for domestic and industrial use, which clashes with demands in agriculture.

It is particularly hard to bring water and lavatory services to impoverished locales and the underprivileged people. Many people in these locales reside without connection to safe drinking water and appropriate cleanliness. Even where enough water supplies are accessible, cleanliness, and disposal of wastewater are often unacceptable. Pit lavatories and septic vessels are the usual means for disposal of human waste but they have narrowed volume and are not always enough to contend with the volume of waste created by many people residing near together. Flooded lavatories and septic tanks adulterate surface water and make a somber health danger (Drakakis-Smith, 2011). The inadequacy of these critical services pressures not only the health and the people environment in slum locales but also that of people residing in orderly urban locales. In Asia and Africa, most of the metropolitan centers have no drainage at all, which affects the upper class and destitute alike. This is authentic of many cities with a million or more residents, as well as tinier towns and cities.

7.6.3. Pollution and Wastes

Urbanization influences water, land, wildlife, and air due to the people number, the volume of buildings and development, and the risen demands on assets. It has collisions on the physical environs in various ways:

1. **Quality of Water:** In advancing nations, many rivers in metropolitan locales are more like open drainage. The inadequate sanitation and drainage systems have an exciting impulse on urban waterways. People use the rivers to predispose all their exhausts from industries commercial enterprises and homes. Wastewater from human residence includes natural material and minerals; industrial effluents include many different typologies of poisonous contaminants. These make the water dangerous for humans to consume for many reasons including irrigation and drinking, as well as killing the fish and more animals and plants residing in the waters. Any transitions to the character of surface water also change groundwater due to they are connected by the systems of the cycle of water so contaminants from the surface will penetrate down and pollute soil and groundwater as well (Figure 7.6).

Figure 7.6. Polluted urban river contains wastes from many sources.

Source: https://www.propertyguru.com.my/property-news/2016/6/129119/polluted-river-to-transform-into-gem-of-johor-bahru-city.

2. **Wastes of Solids:** In many cities and towns, solid waste authority is insufficient or absent. Solid waste plans mean the properly

collecting, transferring, recycling, and disposing of all the solid material we throw away, not excluding paper, plastics cardboard, electrical waste and, food wastes, etc. It also includes hospital, institutional, and industrial wastes which often enclose microbes as well as poisonous and dangerous chemicals, which require exceptional care.

The burning of solid waste brings yet another environmental challenge (Figure 7.7).

Figure 7.7. Waste burning in an urban area.

Source: https://wasteaid.org/waste-sustainable-development-goals/.

3. **Quality of Air:** The quality of air in cities and towns is often very destitute as an outcome of an air contaminant from many diverse origins. These include: exhausts from vehicle, domestic smoke from fires, factory chimneys outputs, diesel-powered engines, construction works dust and streets of city. Destitute quality of air has an important impulse on the health of many urban inhabitants as well as leaving a detrimental and unseen coating of dust on buildings, plants, and other surfaces (Figure 7.8)

Figure 7.8. The brown color of the air adjoining the ground indicates how contaminated the air in major cities can be.

Source: https://www.open.edu/openlearncreate/mod/oucontent/view.php?id=79940&printable=1.

7.6.4. Well-Being

Urbanization can have both confident and adverse effects on health. The major advantages are related to simple connection to clinics, hospitals, and health services in common. If one resides near to these services, one can reach a physician in minutes rather than days or hours, so this advances the necessity to care and common health. There are also goodies from an easier approach to the origins of data such as television and radio which may be used to broadcast information about health to the public. For instance, women residing in cities and towns are more probable to be learned about the planning of family, which results in a decline in the size of family and less repeated childbirth, with the resultant helps to accepted health (Drakakis-Smith, 2011).

7.6.5. Foodstuff

Population migrations also put the burden on supplies of food and the distribution of food. As people move to the urban, they tend to use bought food rather than their crops and this makes them more defenseless to transitions in prices of food. As the population advances and the appeal for water and land accelerate, it becomes hard to accelerate the production of food sustainably. The rise in urban appeal, combined with an agricultural land loss, means an extra burden on rural people to produce food for the advancing number of urbanites. Furthermore, contamination from urban locales can alter the supply of food. For instance, fisheries are often destroyed by urban domestic wastes and effluents of liquid from industries that are city-based. In addition to this, human activities, pollution and environmental degradation which trigger the present climate change had cause decline in food stuff variety for the past one hundred years (Figure 7.9).

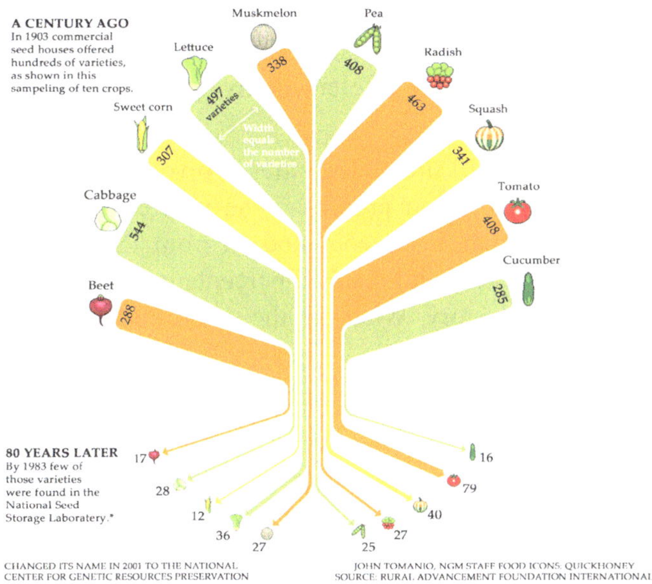

Figure 7.9. A Declining Food Variety.

Source: http://ngm.nationalgeographic.com/2011/07/food-ark/food-variety-graphic

7.7. GROWTH OF MODERN CITIES

Some of the major factors that have led to the advancement of cities are excess resources, automation, and commercialization, advancement of transport and communication, commercial draws of the city, academic, and sporting facilities.

"Like the source of sophistication itself, the source of the city is lost in the vagueness of the past," said Halbert and Gist. There has been sophistication for those who moved from the village to the city.

The first cities seem to have arisen late betwixt 6000–5000 B.C. These cities were nevertheless tiny and hard outstanding from the backyard. By 3000 B.C., there was in continuation of what may be named "true" urban. After that, there was a pause, for approximately 2000 years. It was not before the Greco-Roman periods that cities came into reality. It is interesting that the cities in the locales where city life had sourced finally went into concealment and cities arose in fresh locales. After someday, majority of the objectives of cities in Mesopotamia, India, Egypt, Greece, Rome, and Persia had all been deceived to an elementally agricultural economy (Birch and Wachter, 2011).

In the Western part of Europe, the cities evolved to more abundant and the progress of cities kept on progress. The nineteenth century was the duration of accurate urban insurgency and all from 1800 urbanization has gone forward much rapidly and reached percentages far bigger than at any earlier time in global history. What are the determinants which led to the advancement of cities?

7.7.1. Excess Resources

Cities advance where a community or an association inside it benefits mastery over resources bigger than are basic for the mere nourishment of life. In the old times, these assets were obtained through the occupation of man by man. Forced labor, taxation, or slavery, by the finding or upper class, provided the groundwork for the gain of city life. In current times, man has won over the disposition and lengthened his power. He has used the natural resources to such a bigger amount across technological bettering that now relatively few people can supply the elemental requirements of many. The enlargement of man's authority over nature, certainly in the western nations, has been the basic circumstance of the current rise of cities and the population of the city (Elmqvist et al., 2013).

7.7.2. Automation and Commercialization

The urban improvement has also been majorly aroused by the fresh methods of production-related to modern innovation. The revolution of machinery, the advancement of energy power, and the operation of bigger capital in modern businesses led to the formulation of enormous manufacturing plants which delivered about the flexibility of immobile association of workers accelerating their consolidation around a factory locale.

For the well-being of running with others and of huge wages, men left rural work and current into the modern cities. Therefore, Jamshedpur, a steel center in Chicago India Manchester, Glasgow, Liverpool, evolved into the huge modern cities of the globe. With the anticipation of mechanical energy, a new geographical change has been fashioned.

Already, a collection of peoples was established along the valleys of the river, where the land was abundant and flat. But modern they are erected near the origins of iron and coal. The use of scientific techniques and machinery compelled by electricity or the ignition engine in the manufacturing of goods has now empowered 25% of the population to boost the other 75%, although a century ago 75% was needed to feed 25%. Cities now advance outside much remark to the lands of agriculture (Elmqvist et al., 2013).

While industrialization has aroused city expansion, commerce, and trade also have played an essential component in urban growth. In old civilizations to cities advanced where goods were appropriated and commercial undertakings were carried. Therefore, Sparta, Athens, Pataliputra, and Venice were bigger commercial centers.

In contemporary times, the advancement of contemporary marketing organizations and techniques of trade has greatly devoted to the advancement of cities. Currently, face to face economic transactions require not to be transported in big cities but the mere fact that a bigger proportion of their inhabitants are devoted to "paper" businesses is a meaningful aspect to add to the city progress.

7.7.3. Transport Advancement and Communication

The advancement in techniques of conveyance and connection and the aptitudes which cities give for fulfilling the want for communication also describe the growth of urban. Industrialization relies heavily upon transportation so that raw material and produced goods can be ferried in large sizes. In an industrial city, the means of transport and communication are advanced (Figure 7.10).

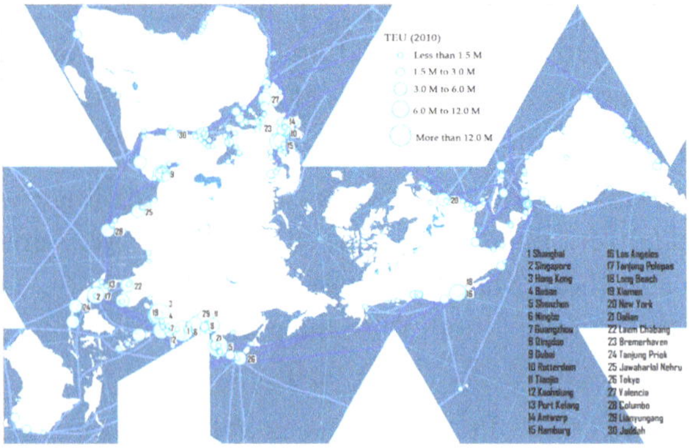

Figure 7.10. World's Major Container Ports, 2010.

Source: https://transportgeography.org/?page_id=9249.

The city is linked not only with other parts in and external of the nation but through advanced means of local ferrying, the diverse parts of the city as well are linked to each other. At the time the factory was established, local ferrying facilities were destitute. The factory employees were forced to reside near their employment place. Overcrowding of housing came.

The local transport supplemented to the city population by prolonging its barriers. The city was partitioned into diverse localities, a market locality, a dwelling locale, a slum locale, and a factory locale and so on. In old cities, inadequate local transportation barred such a calling off of natural locales. The current city is a society that has become overly diversified in its parts (Elmqvist et al., 2013).

7.7.4. The Commercial Draw of the City

Cities make it accessible for more chances for personal growth than do the rural locales. Current enterprises and commerce draw the young generation to the urban where they are paid sufficient incomes. People live in urban not due to they like them as the locale of living but because they can get employment there. Job opportunities are more in the urban than in the rural.

Even entrepreneurs come to the urban from the rural to the advantage of good chances for making higher revenues. As the living standards in the nation accelerate, there is an advancing demand for the variety of goods which are provided in and by urban. This risen request means that people can earn income in a better proportion in the urban. It is in the urban that religious, educational or leaders receive distinctive and high identification. In short, the chances of greater advancement and better residence in the urban account for a better deal for urban progression (Benna et al., 2016).

7.7.5. Academic and Entertainment Facilities

Until currently, all high schools were in urban-like in India. The principle schools in a city are well provided than in rural. Most practice colleges, technical schools, and schools are cited. The majority of big libraries are located in urban.

Examination hubs for aggressive tests are situated in urban and the enrollment agencies are also urban situated. Galleries of Art and museums are in the cities. Outstanding educationists give their lectures in urban. Generally, in the detail of all these facilities young generation are drawn to the urban for higher education.

Facilities concerning Recreation are available in urban. Entertainment theatres and music are in the city. By making requests to the sensation and play the notion of children and adults' same, way they pull them to the urban (Benna et al., 2016).

7.8. CHALLENGES OF URBANIZATION

With transitions in the land-use arrangement when the city rises in size, it grows both horizontally and vertically. The horizontal expansion absorbed the surrounding trimming villages and transformed the agricultural lands so that there is a decline in the level of water. So, there are probabilities of pollution of drinking water due to leakage of pipes. Another thing worth noting is the land value which is accelerated due to deficiency of land in the advancing urban centers. Therefore, there is a sprout rise of apartments and in busy centers; the residents are given consent without checking the way of drainage facilities. The dangers of urbanization may be seen on a variety of lines in the pursuing components.

7.8.1. Ghetto and Related Problems

The strong shortage of housing facilities is one of the most somber challenges disturbing most cities, whether it is an urban city or a minute town. The objective for this is that the opening and advancement of housing facility has not risen rapid enough to meet exceeding demand for fast urbanization arrangement. The serious shortage of accommodation facilities forces the destitute to live in slums. Slums have advanced in almost all global cities. The slums or Bustees have been described by the Indian government under the Slum Area Act of 1954 as dominantly a residential locale, were living because of ruin, congestion, incorrect arrangement, and inadequate ventilation, sanitary or light facilities or any composition of these aspects adverse to health, morals, and safety. It is projected that 40% of people in vast-cities like Delhi Calcutta and Mumbai live in ghettos. These slums have severe filthy conditions. They have poor toilets made by digging a trivial pit in-betwixt 3–4 huts and with sackcloth "drapery" dangling in front. The children, of course, are used to pass anywhere across the huts. All such locales have few pits and pools. These are constantly dug in the central point of a nation's unclean pool. People clean their clothes and utensils below the hand pumps. This leads to diseases like diarrhea, blood dysentery, typhoid, malaria, jaundice conjunctivitis, and, which ambush them all year-round. Children with swollen stomachs or starving skeletons deteriorate from general sight and polio. Human advancement is also severely affected by environmental degradation. Approach to safe water for drinking and dissolution are firmly joined with the expectancy of life and mortality of infant which are very key signs of Human advancement. So, the destitute take decisions on fertility to atone for these factors and to avert dangers. Larger populace leads to more bankruptcy and decays the environment making a vicious circle (Drakakis-Smith, 2011).

7.8.2. System of Transport

There are 300 million trucks, buses, and cars all over the globe. During the crest hours, there will be big traffic jams in the major connections. Due to traffic jams, more products of petroleum are emancipated which results in fuel challenge. During crest periods, the vehicles are parked and overwhelmed and there are more opportunities to happening accidents. The burning of products of petroleum, diesel leads to the increment of carbon dioxide which aids in accelerating global warming, air contamination, and noise deterioration, besides carbon dioxide, carbon monoxide which is freed by automobile. The noise deterioration affects both hearing and non-hearing organs. The hearing effects are deafness and fatigue in human beings. The non-hearing effects are an intervention in communication, speed, annoyance, loss of working adeptness and psycho-physiological clutters.

7.8.3. Garbage Problem

Urban solid waste includes materials for building, hospital wastes, kitchen waste, plastic containers, etc. The materials of building and homes solid wastes are dumped on the communal locales. The hospital wastes do not have wrappings while in shipment. The awful smell pollutes the air. Urban drainage does not have the proper let-out ability. As most society's success, it trashes majorly poisonous metals, plastics, and packing is rising rapidly. In the past decade, refuse was produced at almost twice the population growth rate (Drakakis-Smith, 2011).

1. **Drainage Problems:** The most urban locales in the globe are disturbed with disorganized and insufficient communal amenities. Most cities in the world are not fully sewerage. The objective of this is that the illegal constructions in and about the city lay outside the acumen of the major systems. It has been projected that only 38% of the urban populace has a drainage system. For instance, Mumbai's falling sewer system is a century old, put in place by the British designer when the city was no more than an array of villages of fishing. Currently, it breaks down often with waste about 8 million more people than it was planned for.
2. **Supply of Water:** Some nations have arrived at a stage where no city has a supply of water round the clock. Consistently supply outcomes in a vacuum being made in void lines of water which often absorb in contaminants through leaking joints. Rajkot Chennai, Wadhwan, and Hyderabad obtain water from municipal origins for less than 30 minutes every alternate day. Many minute towns have no major supply of water and rely on such beginnings as household open wells, individual wells, or even the rivers which have some depository water in pools during summer. In the city, the drainage arrangement hardly endures and the yearly flooding of large locales, even in Delhi, it is now a common anomaly in many centers of urban. Mumbai is situated in a depression of keel-line, which also occurs to be the main railway channel. With every monsoon rain, it gets flooded blocking the communication. The problem is specific serious in the Indo-Gangetic plain cities. The situation is bad in the eastern Patna, which remains waterlogged across the monsoon duration. The destination case is that of Katihar were due to the unique bowl-like arrangement of the city and the non-existence of drainage arrangement (Drakakis-Smith, 2011).

7.8.4. Environmental Challenges

Environmental contamination is causing worry and affecting the health of humans currently than yester decades. It has been said by the World Bank that 40,000 people perish in India annually due to air contamination. Current studies also have shown that a vast number of people have been discomfort from breathing diseases cough, and allergies. It has been increased since the 1990s. Further, it has been noted that 23 Indian cities have cruised the hazardous limits due to auto-exhausts and emission from industries. Thus, it is not the work of a central contamination governing board that has to take control but it is the duty of the institutions, individuals to begin possible care and measures to avert the contaminating works. Therefore, it should begin in the form of a communal movement.

8

POPULATION AND DEVELOPMENT

CONTENTS

8.1. INTRODUCTION

Globally, the rate of people being added amongst the existing population is increasing every year. The issue of population is not a mere problem on increased numbers. Instead, it is an issue of human welfare and it affects development. Past studies have concluded that when population growth is rapid there are some serious repercussions for the well-being of humanity globally. The term development has had many definitions. In this chapter, we will refer to it as the improvement of the level of living of people's life, which is their health, incomes, education, and the overall well-being. It also entails their respect, dignity, self-esteem, and freedom of choice. Population growth rate has been found to have a relationship with development for example in Africa this is a major problem but not the only factor. There are other many countries globally and not just African countries which have also witnessed the impact that population growth rate has on the development aspects.

There are major issues that affect population and development such as the level of living, increased in labor force, unemployment, poverty alleviation, improved education, and health, freedom of choice, etc. The level of living for a population is anticipated by the growth rate of the population since it has a direct impact on the demand on the social services such as housing, transport, security, and sanitation. The level of unemployment also tends to rise in high populations due to increased labor force. It is crucial to weigh the employment opportunities available and how the unemployment levels should be managed. Poverty alleviation levels can have huge implication to the poor people due to lack of sufficient reach to resources that are needed for an adequate diet. In regards to education and health, high population also led to a strain in the health and education systems. Freedom of choice and poverty are also resultants of high populations at the family unit level due to the size of the family. The densities, numbers, and rates on the available opportunities when the population growth rate is high are also another factor that shows there is a relationship between population and development. In the current chapter, we will exploit how population contributes to the realization if development (Gould, Population, and Development, 2015).

8.2. UNDERSTANDING POPULATION GROWTH RATES AND DEVELOPMENT

The population at a worldwide scale is currently (2019–2020) rising at rate estimated to be 1.08% annually. Globally, the average population is known to increase at 82 million per year. In the 1960s, the annual growth rate was at its peak at estimated to be around 2% in the late years of that decade. According to research, the rate has rose by half since then but it will continue to decline over the years. In the 21st century, the world population rate is expected to rise but at a slower rate compared to the recent past. Between the year 1959 and 1999, in a period of 40 years the world population doubled. It is estimated that by 2037 the population will be at 9 billion which is a 50% increase over 40 years.

Population growth in developing countries is another observable trend. The birth rates in underdeveloped countries nowadays are significantly greater than they were in pre-industrial Western Europe. From 1 750 to 1850 their population grew at an annual rate of 0.4%; from 1 850 to 1950 the rate was 0.9% while from 1950 to 1975, it had risen to 2.3%. In 1960, about two-thirds of the World's over 3 billion people lived in the developing countries. In the 1970s, the annual growth rate of population in these countries was even higher; in Latin America and Africa, it was at 2.7 and 2.6% respectively while for Asia it stood at 2.4%. In the 1980s, the growth rate was higher still because, by January 1982, the World population had passed the 4.5 billion mark while by July 1987 it had hit the 5 billion mark. In the year 2000, earlier scholars predicted the World's population would consist of over 6 billion people. The most unfortunate thing about it all is that the developing countries would have over 80% of the projected total. It was also speculated that in the year 2000, about 77% of the projected population would be living in the developing world (Gould, Population, and Development, 2015).

In the current section, the rapid population growth is witnessed at the rates above 2% and this is especially common in the developing countries. This alarming growth rate is seen as the bottleneck to the development. The population growth can be billeted up to a certain point but beyond that point, it proves to be gruel some. Over the past three decades, there are many countries that have been able to raise the average income despite their population having a rapid growth rate. In a strict sense, some countries have been able to accommodate population growth. The goal of development spreads beyond the accommodation of ever-larger populations since it aims at improving the lives of the people. The population growth rate, particularly in the developing countries has ensued in less development that might have been a different case in the raise of the living standards amongst the large numbers of poor people in the world. It is clear-cut that the rapid population growth results in a slowed development rate (Dyson, 2013) (Figure 8.1).

Figure 8.1. Population pressure and development are broad concepts that have been studied by various scholars, economists, planners, sociologists, geographers, and demographers for different reasons.

Source: https://villageenterprise.org/what-we-do/what-we-do_extreme-poverty2/.

Population as a factor influencing the levels of living of people has been part and parcel of society from ancient times. Several scholars have accepted the fact that the average condition of people deteriorates if population continues to grow beyond the available resources. A rapidly expanding population can be an economic burden if it does not balance with the growth of the economy. The population reduces output per work and depresses the level of living of the masses. This idea was also shared by Plato and Aristotle who said that man's potential could be fully developed and his highest good realized if there was an optimum size of population, not exceeding the available resources.

Accordingly, poverty had its origin in the population pressure on subsistence. The size and growth rate of population is a major obstacle to the improvement of living conditions of the people which results to the negative points of population pressure. However, the natural law through war, disease, and starvation would inevitably take care of population pressure especially in the earlier years. However, the current article attempts to give feasible planning recommendations to counteract population pressure and not through a natural process as advocated by earlier researchers (Dyson,

2013).

The growing population pressure on resources of the world through sheer number of people and overcrowding on the land and other limited resources can only be used for human fulfillment (which encompasses t material, space, beauty, recreation, and enjoyment) if some balance between the population and resources exist, thus increasing food production and controlling birth. An uncontrolled continuous population pressure results into increasingly desperate problems. For a long time, the population issues in Africa have been an interest for many scholars. The major factors that affect population can be said to include independence, fertility, and education which are keys determinants to development.

Learning from the earlier days, the solutions could be effective if fitted into socio-economic planning and their administration made part of the normal plan implementation process. The anticipated constraints are limited resources, incomplete data, and other cultural and institutional parameters. To tackle the problem of overpopulation and internal migration it is important to consider the impact other parameters such as land shortage, unemployment, underemployment, low standards of infrastructural facilities, ineffective extension services, and low standards of formal education affect the realization of development. Past studies indicate that the underdeveloped world is facing a deteriorating population-resource balance; there is a limited resource base (Dyson, 2013) (Figure 8.2).

Figure 8.2. The rapid population growth penetrates into all spheres of a nation's economic and social life; and that they are instrumental in shaping such things as the man to land relationship and the structure of the community.

Source: https://www.thinglink.com/scene/551036071878590465.

Overall, while a new pair of hands accompanies every new mouth that comes into the world, it is only under certain circumstances that the new pair of hands would produce as much food as is produced on average by those pairs of hands already in existence and at work resulting to unbalanced equilibrium that affects level of income and livelihood. The next section will discuss population and sustainable development.

8.3. POPULATION AND SUSTAINABLE DEVELOPMENT

Sustainable development relies on three main aspects that is economic, environmental, and social sustainability (Figure 8.3).

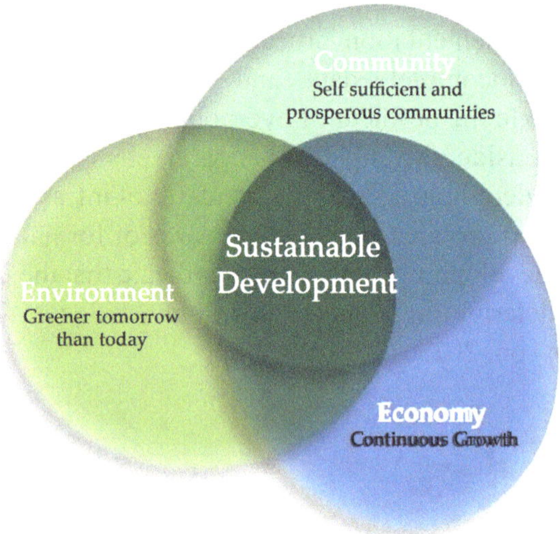

Figure 8.3. Aspects of sustainable development.

Source: https://www.kullabs.com/classes/subjects/units/lessons/notes/note-detail/1137.

From the previous section, the low level of income equilibrium is one aspect categorized in the economic pillar. It is the result of population growth. It is the opinion of this article that whereas there have been improvements in technology, most of the countries in Africa, Asia, and Latin America is yet to fully reaps from it.

Although the human population is growing fast, the round earth on which we live is fixed in size, and that as more people crowd together on it may seems to grow smaller all the time. The population growth is a threat to the availability of food, land, and water. Populations are growing so fast in the Third World countries that they are finding it difficult in providing enough food, houses, schools, and jobs. The governments of these nations spend so much money just trying to feed, house, and educate their fast-growing populations that they have little money left over for building new roads or for venturing into income-generating investments. It is in this light the population growth has been seen as an impeder of economic development besides being a strainer of the natural resources (Gould, Population, and Development, 2015) (Figure 8.4).

Figure 8.4. Population growth is currently one of the most serious threats to human happiness and progress and that it is a threat to all our efforts to raise our living standards.

Source: https://citizentv.co.ke/news/africas-rapid-population-growth-puts-poverty-progress-risk-gates-212363/.

Some scholars argue that unless population grows is restrained, we may have to abandon for this generation our hopes of economic progress. Progress may not be feasible with the current rate of population increase. Rapid population growth slows down the growth of per capita incomes in less developed countries and tends to perpetuate inequalities of income distribution. It holds down the level of savings and capital investment in the means of production thereby limiting the growth rate of gross national product.

In total agreement with the foregoing, past scholars observed that countries which are now in the developed category, population growth and economic growth usually went together but an opposite in developing countries. In these countries, rapid population growth increases the dependency burden thereby reducing the ability of the working population to save. This, in turn, makes it difficult to create the necessary capital to productively employ the growing population. The picture which emerges here is that population growth perpetuates the vicious cycle of poverty. In the long-run, the course of economic development is inextricably bound up with changes in the size and composition of the population. In addition to the foregoing, many statements have been made by leading personalities on the population question. One reason for this is their realization that population growth is one factor which is capable of thwarting development (Gould, Population, and Development, 2015).

Accordingly, mankind's history points to the fact that the depressive economic and human miseries resulting from too many people and too little food and economic opportunity in one part of the world undoubtedly imperils the security and well-being of another. It is against this background that it has been said that widespread poverty, low productivity of labor, the growing demands for food, and slow industrialization distort and degrade the international trade mechanism. In other words, it sees population growth as a global crisis which calls for international cooperation in seeking for a solution. It is, in addition, observed that large scale internal migration and rapid urbanization are among the most important social effects of rapid population growth.

Furthermore, the growing numbers of children who survive their parents place new strains on intergenerational relationships. It has been pointed out that population pressure results when humans and their activities exceed the carrying capacity of a finite region to feed, clothe, and otherwise sustain them. It is in this light in some parts of tropical Africa, there is already real population pressure on land. The fact that many of the Third World countries have not yet "settled" substantially in the industrial sector further negates their prospects of quick economic progress. Most of their people are born having a little choice to make as to what their source of livelihood would be. They turn to land and take agriculture as their occupation. The problem with this as the following section would show is that land is an exhaustible commodity which cannot perpetually bear the pressure created on it (Hawthorn, 2013).

Internal migration has been the most extensively studied component of population dynamics and certainly the one often considered in the context of a wide range of socio-economic phenomena which influence it and are in turn influenced by it. The three components of population dynamics, namely, fertility, mortality, and migration are crucial in development studies because they influence trends in population structure as well as change. Migration, the most widely studied of the trio, has attracted a broad, and consistently expanding group of scholars in social, physical, and biological sciences. Geographers are particularly concerned with spatio-temporal analysis of migration investigating among other things the matrix of determinants, a host of consequences and ingredients of the process itself. Their broad approach is instrumental in incisive analysis adopted by demographers at a more refined level. For example, the present pattern of population in modem Africa is a function of population movements that are deeply rooted in pre-colonial times but that have persisted through the colonial era to the post-colonial period, it is important that it be explained in spatio-temporal' perspective.

In theory, population projection is a relatively simple calculating operation based on an assumption about trends in fertility, mortality, and migration. They may be prepared for the total population of a nation, its principal geographical subdivisions, or specific localities within it. Further, they depend on the availability of accurate population data for a base period distributed by age and sex. The principal uses of population projections relate to economic and social planning. Economic and social planning is a crucial element in the realization of sustainable development. Taken together, these plans constitute public policy related to accelerating social and economic development. The first step in planning is to study relevant aspects of the population and the economy, both at the present time and in the recent past. The next section will review the new understanding on population and sustainable development based on the result of six major international conferences convened by the United Nations (UN) during the 1990s (Hawthorn, 2013).

8.4. UNITED NATIONS (UN) INTERNATIONAL CONFERENCES 1990'S

The rise of new understanding on population and sustainable development under this meeting stressed on a global, national, and local organization of the UN resources to look at the way the environment is hugely affected by human activities and thereafter seek a suitable solution focused towards sustainability.

8.4.1. International Conference on Population and Development - Cairo, 1994

In this conference, the UN discussed the impact of population on the economic and social development and emphasized the role and interests of women. Matters pertaining to sexual and reproductive health rights were emphasized on and it was concluded that a new meaning of population policy towards prominence reproductive health and the empowerment of women was necessary.

8.4.2. The World Summit for Social Development - Copenhagen, 1995

In this Summit, the objective was to found a people-centered framework in the aspects of social development and building a partnership/cooperation culture. It was stressed that it was vital to address the immediate needs of the human in distress that were affected.

8.4.3. Fourth World Conference on Women - Beijing, 1995

This conference was centered on the need to take action for equality, development, and peace. The conference marked a significant turning point for the global agenda for gender equality. It recognizes that equality of women and men is necessary for social justice and to achieve development and peace. It is an action agenda to safeguard women's human rights. The goal of the platform is to accelerate the implementation of the Nairobi Forward-looking Strategies for the Advancement of Women, which will succeed only by commitment of Governments, non-governmental organizations (NGOs), and the private sector, equal participation of women and men, and strengthening of mechanisms for accountability to the world's women.

8.4.4. United Nations (UN) Conference on Human Settlements (Habitat II) - Istanbul, 1996

The aim of this conference was to address the worldwide importance on sustainable human settlements in the urbanizing world and the need for adequate shelter for all. Human being is placed at the center of concerns and they need sustainable home settlements, adequate shelter, healthy, and productive lives.

8.4.5. World Food Summit - Rome, 1996

This summit is in response to the sustained existence of prevalent undernutrition and a growing concern with regards to the capacity of agriculture to address future food needs. All people have the right to free from malnutrition and hunger in an effort to developmental and physical faculties.

Overall, education is associated with the better social development. It results in fewer complications among pregnant women if better-educated women are in general in better health than others before and during pregnancy. Education could be associated with greater likelihood of receiving appropriate care for complications that do arise. Women that are more educated might also be better informed about the symptoms of complications and could therefore be more likely to make a timely decision to seek health care when a complication arises. Such women might also be concentrated in urban areas and thus could live closer to health facilities, or they might have better access to transportation needed to reach those facilities.

Finally, educated women might be more likely to receive appropriate and timely care either when they reach a health facility, because they are better able to pay for the care or because, by virtue of their status, they are more likely to be well treated. In regards to poverty and malnutrition, it is important that a healthy nation. In this article, the analysis, poverty is stated in relative terms. It refers to a household that fails to satisfy the requirement of adequate food, a household that stays in slums and includes street children and those who fail to get access to adequate necessities like water and shelter. Relative definitions of poverty consider not only the ability to consume a certain minimum of goods and services but also compares the poor with those who are in other income brackets (Hawthorn, 2013).

Increasing levels of poverty in urban areas would have far-reaching planning implications. It would have implications on the maintenance of law and order, development of informal housing or squatter settlements, environmental deterioration, and on general urban land use policy. A measure of poverty helps planners to know how many people are poor at a certain period in time. This will guide towards a clear policy on the provision certain basic needs like affordable housing or allocation of land. Non-governmental organizations and other charitable organizations by looking at trends in the poverty rates over time can determine what is happening to the numbers of the poor population and try to mobilize community resources to help improve their welfare. Any policy to tackle the problems of poverty must take into account the causes of poverty. There are two main views about the causes of poverty. The first view blames poverty on the management of the national economy while the second view blames poverty on the individual. The premise for these views is that some people are poor because they have made bad choices while others are poor because of forces beyond their control.

Most poor people have little education; they would thus require certain forms of employment in urban areas. This can lead for the formulation public works programs like say food for work. To some extent, many poor people actually work (the working poor). The main cause of poverty for this group is low wages. For this group, employers may be encouraged to build affordable housing so as to enable them spend less and afford basic food to maintain and increase labor productivity. A suitable means of transportation that is relatively cheaper like urban railway network or non-motorized transport for this group can be planned for. The cost of living is generally higher in urban areas partly because of the fact that all food to be consumed is bought as compared with households in rural areas that can produce some food from their farm (Hawthorn, 2013).

Food accessibility levels influence household consumption patterns. More research should be done at a micro-scale level to determine levels of food security and accessibility and its implications At the moment research on food accessibility is done at the macro-scale (National level) are the total harvest of food is compared with the total population. The Provision and improving access to infrastructural services to the urban poor is an important and an essential ingredient of poverty reduction that will consequently their standards of living. For instance, the provision of piped water to the urban poor would reduce household expenditures thus leading to the improvement of food accessibility and health conditions the households.

8.5. STRATEGIES DEVELOPED TO ADDRESS THE ENVIRONMENTAL AND DEVELOPMENT PROBLEMS CAUSED BY POPULATION ISSUES

There are a number of strategies that are applicable in addressing the development and environmental hindrances that are caused by the rise in population. They include:

8.5.1. Slowing Population Growth

The average household size persons in a family should be regulated to make sure it is not very high given the resources available. Solutions such as reducing the household sizes have revealed the need for family planning programs. Family planning programs were initiated by the different governments thus advocating the spacing and control of children. The Ministry of Health in liaison with organizations such as UNICEF has been at the forefront on family planning issues. However, in some countries, cultural hindrances have constrained these efforts. For example in some African cultures, the cultural constraint springs from the fact that the people believe God gives children and is responsible for their sustenance no matter what the number. Consequently, it is a delicate issue that needs to be communicated carefully especially to the people of low formal education globally. Despite the constraints, family planning campaigns need to be intensified and integrated into the people's culture.

The local people should be made to appreciate family planning from the grass-root level. It becomes significant and necessary to recruit indigenous family planning educators and other charismatic leaders from organizations who have accepted family planning. These are the appropriate family planning agents. The failure of implementing the reduction of birth rate is not due to lack of effort, however, it's due to poor communication. There are problems in approach in the sense that people with large families acre depicted as social failures through films. They can do nothing about their large families; consequently, they are alienated and thus completely ignore issues of family planning. Given a better approach, these are the people who would convince the younger generation to have fewer children after being made to understand the problems of having too many children beyond one's capacity to manage and care. On the overall, the family planning educators on their part should be trained sufficiently in public relations techniques to convince and mobilize the people in accepting the campaign for family planning (Simon, 2014).

8.5.2. Building North-South Partnerships

There are increased efforts at a global scale to build partnerships for North and South which have a global significance in the contribution of more research capacity in low- and middle-income countries (LMICs) over the historical few periods. The lack of sufficiently skilled researchers to appraise health policy development and this is evident especially in LMICs. Projects such as the EU FP7 funded African/Asian Regional Capacity Development (ARCADE) have been at the forefront of the development for a new generation of well-trained researchers globally from universities. This will help in the focus on health-related subjects, services, and social aspects (Simon, 2014).

8.5.3. Ending Poverty

Educating human beings to be the most productive is a significant element of development. This involves the distribution of education facilities of all levels. Health facilities are also a necessary component of development complimentary to educational facilities in necessitating productive human beings for the efficient utilization of resources.

To overcome environmental and development problems, there is a need to build more health facilities to serve the Southern and Western parts of the division which are deficient in the health services. These services need to be located in service centers with improved road network; the population would have easy access to these services. These services should be coupled with increased training of skilled manpower to man the institutions. The health facilities expose a poor spatial distribution of the services which are located in the less developed countries. There is a need to have health facilities distributed evenly all over the division to serve the area efficiently.

To reduce poverty, another aspect is the provision of adequate and efficient infrastructure is necessary for development, as they form part of the significant interdependent sectors that aid the economy functions broadly. The infrastructures considered here are communication, water supply, and energy. Communication is another factor and it refers to several means that enable exchange, transportation of messages, information, people, and goods ranging from mass media, postal services, and technical transport modes. A developed, efficient, adequate, and reliable transportation system is necessary for the development and it is necessary especially in less developed countries that have ignored this aspect for many years. The transportation network is significant as it connects areas in a region for varied functions.

8.5.4. Creating a Favorable Economic Climate

According to the overall model for a favorable economic climate, it is important for countries to have favorable factors such as political stability. For suitable growth to take place, the basic industries must expand for this brings more income and a new population to the town centers, thereby increasing the demand for local services. This is especially common in Africa and Asian developing countries. The non-basic industries also expand, further increasing the population's purchasing power and demand. In the context of the theory, basic or town-forming functions can be defined as those activities which bring revenue to a town by serving an external market. On the other hand, non-basic or town-serving involve the provision of those goods and services that a town's own population needs. The growth of urban communities is the result of an interplay of many forces. It is the product of many decisions by the people who live and work in our urban places. Urban growth is caused by decisions such as those to move the family or decisions to seek a new and better job or a more interesting social life (Simon, 2014).

On the overall, transportation is identified as a major problem in the developing countries due to the poor conditions of the roads which have partly resulted in the absence of regular services for means of transport. This results in predominance of walking as the mode of transport for the areas. Walking as a mode of transport is neither adequate nor efficient in transporting products for marketing. In fact, in agricultural areas, the road network is necessary for transporting the cash crops and other products for sale and carrying consumer goods from the manufacturing centers to the people. Currently, the problem of poorly maintained roads is a major issue.

8.6. ENVIRONMENTAL IMPACTS OF POPULATION

The population affects the environment resources. This section will review the different aspects that are affected.

8.6.1. Land

Land has been an invaluable treasure to man throughout human history. In spite of this, however, it is not a commodity which can significantly be increased to offset its ever-increasing demand.

In most developing countries, out-migration to the prime centers is only a temporary solution to the problem of pressure on land, as the migrants still have a claim to the land in the rural areas. The manifold effects of population growth penetrate into all spheres of a nation's economic and social life; and that they are instrumental in shaping such things as the man to land relationship and the structure of the community. Although the human population is growing fast, the round earth on which we live is fixed in size, and that as more and more people crowd together on it, it seems to grow smaller all the time. Her conclusion is that population growth is a threat to the availability of food, land, and water.

8.6.2. Water

Water is the basic source of water for human, livestock, and other activities for most of the families. Currently, it is a national goal to provide everybody in the rural areas with piped water sufficient for human and livestock needs in the developing countries. While the incidence of droughts and water-borne disease in the division is not acute, there is the problem of distance traveled to fetch water. Similarly, water supply is inefficient as the majority of the families depend on untreated water. Consequently, there are some water schemes in the division meant to improve on the situation of water sources and supplies.

8.6.3. Air

The increased reliance on traditional sources for energy such as Wood has led to deforestation consequently affects the air. Additionally, wood has been used to provide building materials besides being used for heating and cooking. In fact, there is overdependence in the developing countries which has further led to increased environmental pollution. Wood is the main form of energy in the form of firewood or charcoal. However, due to high population densities people have cleared former forest areas for settlement and agricultural, thus limiting the available timber resulting in shortage of energy supply. This has a direct impact on the properties of the air. Additionally urbanization leads to further pollution.

8.6.4. Biodiversity

Rapid population growth has an impact on the biodiversity due to the natural hazards. The growth of the human population has result to a huge impact on the planet and exaggerated biodiversity. The impact for food, raw materials, and space has increased the pollution levels and destruction of habitats.

8.6.5. Energy Consumption

Consequently there is need to provide alternative means of energy. Wood, which is used for cooking and lighting is relatively expensive the environment in developing countries. Oil is also an exhaustible and immutable resource consequently; it's not a solution for energy in the long run. Rural electrification is a feasible alternative source of energy as long as there are rivers that can be utilized for generation of power. There is need for man-power with technical know-how on how to generate electricity given the available resources.

Besides the above forms of energy there is need to consider other means of harnessing all forms and source of energy and thus conserve the energy in the area. This involves a look into the use of biogas and solar energy and where possible afforestation or reforestation to rehabilitate the deforested area. Similarly, energy saving charcoal stoves should be encouraged in the area.

Thus, power is greatly needed if the rural activities are to be diversified. Electricity is significant for large scale industrial activities like paper industries, quarry projects, sawmills, and other rural activities.

8.6.6. Urbanization

Examining factors of urban growth observes that population increase is an important factor in urban growth. Larger cities are the places where real population changes can rapidly occur due to natural increase because the population is already great and particularly because people in the fertile age exist already in reasonable proportions. In a different study, the urbanization in the Middle East observes that the main factors responsible for urban growth in the region were political centralization, internal political strife, changes in foreign trade patterns and foreign relations and the discovery and exploitation of oil.

"Demographic" and "migration transitions" whereby high fertility generally canceled by declining mortality results in a fast natural increase, causing the preponderance of youthful population which is by far the most migratory age-group and engendering considerable rural-urban migration which has been partly responsible for rapid urbanization. This process of urbanization is due mainly to rural-urban migration which is in keeping with the phase of "migration transition" in the developing countries. As spatial physical planning among other things undertakes to create more "growth centers" that would organize regional economy and encourage intra-regional mobility in which human resources would benefit the region of provenance, extra-regional migration streams could be dammed at their origins without any coercive measures for solving the problem.

Overall, it is evident that infrastructure is the backbone on which rural development depends. There are some shortcomings experienced in the developing countries such as poor communication network. Road network is the principal form of transport in most of the developing nations. Generally being agricultural areas most of the areas in developing countries is accessible by road. However, transportation still remains a major problem in the area due to poorly maintained roads, serviced with buses and vehicles. They are impassable during the rainy season and very dusty during the dry season. These poor conditions of the roads have resulted in the absence of regular services. In bid to provide everybody with water, the problems to be considered in this respect are the successful distribution of water to the families and ensure it is done on a constant basis. Finally, energy is the least developed resource in the division and yet the most needed for rural industrial activities (Simon, 2014).

8.7. THE RIGHTS OF PEOPLE TODAY

Today, the Bills of Human Rights are fundamentally the rights of all individuals by virtue of being a human being. There are legal rights which provide protection and they range from the municipal to the international laws. The human rights are universal and are protected under the law. They include:

- **Right to Life:** All people have the inherent right to live and they should not be killed by other person(s);
- **Right to Fair Trial:** All people have the right to have their trial being fair and conducted in an impartial court;
- **Freedom of Thought, Conscience and Religion:** This right protects the freedom of choice to follow the doctrines of a particular religion;

- **Freedom from Slavery:** Slave trade and slavery has been abolished despite it being exhibited in some parts of the world; and
- **Freedom from Torture:** This is prohibited under international laws. There are other many rights and they include:
- Freedom from discrimination;
- Freedom from interference with privacy, family, home, correspondence;
- Freedom of movement;
- Freedom of speech;
- Right of opinion and information;
- Right of peaceful assembly and association;
- Right to adequate living standard;
- Right to education;
- Right to liberty and personal security;
- Right to marriage and family;
- Right to nationality and freedom to change it;
- Right to own property;
- Right to participate in government and in free elections;
- Right to remedy by competent tribunal;
- Right to social order; and
- Right to social security.

In line with the current article topic on population and development, it is impossible to discuss all human rights in the current times. However, a brief review will be given on the rights that relate to the context being discussed. In the context of the current article on population and development, to substantiate, spatial, economic or sociological analyses ‘ of the determinants of migration ultimately produce models that are far from complete but which, in their own right, present varied perspectives of the factors involved. In comparison with today, the past was biased. For example, traditionally land was acquired with no strict defined boundaries through rights by first clearance, rights over large grazing areas, rights through marriages and inheritance, this type of ownership was complex with the clan, lineage, family, and individual holding rights in a decreasing order. The individual rights over land were minimal; however, the land problem was not critical as the existing population did not exceed the available land. It is this sense of kinship loyalty which led to homesteads being nucleated in one particular place as decided by the elder of the clan.

As for Africa, in the traditional setting, land was held in common. Communities were self-reliant and developed a high level of social solidarity and egalitarianism. Land was abundant such that the system governing tenure and control was based on the need for labor to clear the land in order to bring it under cultivation. The person who cleared the land became the owner and the heirs continued to own the land as a group; each family having the inalienable right to enough land to maintain itself. Thus, group ownership and individual usage of land became the norm.

50 years ago, social groups which take on identity of persons continued to be the basis of legal rights of the individual and the community in the African social and economic setting. In the modern sectors of the economy, community associations or groups especially in the urban areas are gradually overcoming the tribal and ethnic influences. In regards to lands, conceptions of Common and African Customary law encourage private landowners to exercise their freedom to pursue economic rights without effective public policy. Furthermore, the freehold land ownership rights which are exercised without effective enforcement of planning regulations are responsible for the lack of concern for the environment.

In the education sector, primary education is a basic human right and a prerequisite for socio-economic development, it is necessary for the nation to ensure that all those in the appropriate ages have access to the necessary educational facilities. This can be done through expansion of the facilities in conjunction with a suitable population program to ensure that the demand for the facilities does not outweigh the capacity of the society to supply.

Focus should be on the primary education as a basic human right and the minimum basic requirement for take-off into the modern sector of national life especially in developing countries. The demand for educational services tends to grow as the population increases, in so far as every human person has a right to be given a chance to develop his mind and skills, and society has a duty to make the best use of available resources.

Women especially in the developing nations have been fighting for equal rights, a step that is seeing continents such as Africa and Asia shift towards overall improvement. The UN recently launched a global campaign aimed at fully mobilizing the untapped economic potential of women globally. This move was an affirmation aimed at providing economic opportunities to women and girls, a right move towards stronger growth and higher levels of their economic prosperity. Overall, human rights have done away with most of the vices in the world today, a factor that has largely led to the realization of social development and other improvements as well. The bill of rights have ensured that oppression to the weak in the society is minimized with a strive to completely eliminating inequality as far as universal rights are concerned.

9

WOMEN AND SOCIAL DEVELOPMENT

CONTENTS

9.1. INTRODUCTION

Women perform an important a role through their participation in social development in the local, regional, national, and global levels. In the current article, social development refers to a wholistic method that entails the economic, cultural, political, human, and technological development of the societies. Although traditionally many aspects of day to day life such as businesses were owned by men, in the last three decades women all over the world began starting and owning their businesses to match their male counterparts. Businesses owned by women worldwide have an admirable success rates. Statistics show that businesses owned by women have been generating impressive income, which has led to increased social development.

9.2. WOMEN EMPOWERMENT

The subject of women empowerment continues to gain traction in countries all over the world. Many organizations across the world are playing different roles and contributing in numerous ways to women empowerment. For instance, some organizations are providing financial facilities while others are offering training and development activities to women (Figure 9.1).

Figure 9.1. Most women entrepreneurs in the world have progressively begun to extend their benefit in the economic sphere by successively owning and managing small and medium enterprises (SMEs).

Source: https://startwheel.org/women-entrepreneurs-gather-in-williamsburg-to-raise-awareness-and-capital/.

Women-owned SMEs especially in Africa face a myriad of challenges and are less profitable due to low levels of education, diversion of funds meant to fund business operations and large family numbers depending on the business income. In developing countries, SMEs are by necessity once one lacks formal employment, majorly because the education system is skewed towards producing graduates ready for formal employment rather than creative agents for establishing entrepreneurial ventures (Boserup, 2007).

Women entrepreneurs experience social and economic discrimination because the society in most developing countries is patriarchal. In developing countries, the performance of SMEs is impressive and the government supports the private business sector in an aim to not only reduce unemployment but also improve economic growth. It is important to support women towards self-employment guided by a good policy framework from the government is a major way that a government can build a successful entrepreneurship population. In terms of economic growth and employment, SMEs have played a critical role in developing countries. Many developing countries growth is largely dependent on SMEs support for job creation and GDP contribution. SMEs exist in every section of the economy from trade, farming, manufacturing, and service provision. Despite the crucial role played by SMEs in supporting this country's economy, women have been left behind until recently. The 21st century has witnessed the glass ceiling effect, which has had an impact on women to overcome hindrances and succeed in undertakings such as businesses (Boserup, 2007).

Empowerment initiatives are a set of activities in which the subject of women empowerment continues to gain traction in countries all over the world, both developed and developing, with focus more slightly on the underdeveloped countries. In the last few decades, there has been a global shift towards women empowerment. Women rise to strengthen social development promotes diversity and inclusive representation of women globally and helps women to market and promote their products by launching a dedicated women-owned product marketplace on its trading website. There are other global business firm that, through its business activities that aim to economically empower women by offering business training and financial counseling with lending services to provide budding individualized technical assistance they need to learn, grow, and succeed.

In Africa, microfinance by granting small loans to poor women entrepreneurs gives them capital used. IT is important to have initiatives that promote women empowerment through specialist training, skills development and assisting women to establish their own enterprises. There are empowerment program in Africa that purposes to lift women economically. Compared to other countries such as Mexico, the performance of women-owned SMEs has been greatly improved by empowerment initiatives initiated by financial institutions. Some of the findings show that the failure to seek formal finance compared to male counterparts is contributors of low performance of women-owned enterprises (Boserup, 2007) (Figure 9.2).

Figure 9.2. Economic potential of women is gaining greater attention in the developing countries hence attracting much policy and literal attention.

Source: https://www.weforum.org/agenda/2016/03/how-can-african-women-achieve-their-economic-potential.

Women are key facilitators of microeconomic development and contribute significantly to the GDP and improving incomes of their households. The support of women empowerment by different government through funds such as the Women Enterprise Fund and the affirmative action for women in the education sector, Nongovernmental organizations, Micro-finance institutions, and women themselves through self-help groups has aided them. If the main objective of economically empowering women and strengthening their access to resources and opportunities is to be realized then their contribution to social development will be cemented. The institution's programs can be broadly categorized into four: financial support, capacity building, marketing support, and creation of business networks (Leach, 2015).

Different sources of credit for small businesses that women can tap include; trade credit, external financing such as bank debt (domestic and foreign). In the encouragement of women participation in social development, finance remains a major challenge for women-owned SMEs as they are often not able to get long-term finance in the form of debt from banks and other financial institutions. Access to finance is difficult and is due to among other factors banks' points of restriction and credit establishments. In some countries especially in the developed world, governments offer financial support women-owned SMEs with the sole goal of accelerating economic growth of their countries. In some parts of Africa, microfinance which entails provision of credit facilities to wealth and exit poverty. The small loans are granted without collateral, as it is understood that poor people from the poor. Women have largely contributed to social development and in their support, it is important to provide formal training and development for women because through the training, they are equipped with basic skills to help meet other social development contribution (Boserup, 2007).

9.3. INCREASED ECONOMIC INDEPENDENCE FOR WOMEN

One of the ways that women have contributed to the economy is through their business roles. In a quest to achieve growth and enhance performance, SMEs owned by women face numerous challenges including marketing of their products and unfortunately, some SMEs are unable to successfully market their products. Women are inclined to help create satisfied customers who in return are the organization's cheapest customers as they buy repeatedly, talk favorably about the business hence free advertising. A report by UNCTAD identified promotion of SMEs' exports to other countries as a vital factor in helping SMEs to attain competitiveness in the local and even global markets. A report by that conference titled 'According to the findings of the research paper, some SMEs have performed below expectation due to marketing challenges. Marketing problems of SMEs impact on the profit-margin as well as the sales volume of SMEs, hence managers of SMEs should be adequately supported on the principles and practice of marketing.

In Africa, various trade blocks especially COMESA and EAC have created business linkages between SMEs and multinationals across the markets. SMEs in countries such as Kenya continue to face limited access to markets owing to globalization. SMEs are less aware of opportunities due to lack of information. To improve market access, most governments through the ICT Ministry enable SME customers to facilitate communication with customers, competitive positioning, and information acquisition and also assists women market their products.

In some developing countries, women are responsible for providing labor for household chores besides agricultural labor, which leads to the feeding of the families (Leach, 2015).

At the global level, new attention is focusing on the glass ceiling effect for even in more developed countries of North America and Europe. Differentials in maternal mortality by socio-economic status exist among countries. However, socio-economic status is a complex concept at community level which affects factors such as their occupation, their level of personal income or wealth, and their autonomy (e.g., their ability to travel on their own or make independent decisions to use health facilities). These are factors at both the macro and micro levels. At the macro level, the economic level of the whole country is usually considered. For example, the article recommends consideration of whether the country has the resources to have a well-spread and easily accessible network of women health, maternity homes, trained midwives, and health care centers. The macro-level variable considers the general educational level of women in the country, which in turn affects the level of understanding of mothers and would-be-mothers, their ability to make use of the said health and/or maternity services and facilities to their advantage.

The world has witnessed increased economic independence for women given their capacity building and performance in all sectors in the economy, which contributes to social development. Capacity building and skill development is critical in enabling women to perform well and enhance their competitiveness globally. It has been established that order to address such barriers, stakeholders such as governments, public, and private sector institutions, and financial institutions should provide technical assistance and capacity building to women entrepreneurs to increase their economic empowerment (Leach, 2015).

Non-governmental organizations also play a major part in capacity building because of their traditional role of gap-filling through basic education. For instance, United Nations Conference on Trade and Development (UNCTAD) in an effort to meet their mandate and become competitive in global markets, identified enhancement of productive capacity through use of technology amongst others as a key component of improving competitiveness of SMEs (UNCTAD, 2005). Globally, the current global business environment is compelling business firms to adjust and adapt to changes in order to continue to exist, grow, and remain competitive. Inter-company co-operation and business networks are some of the strategies being adopted by firms in the current dynamic business environment and women have been on the forefront evidencing the emergence of women business leaders and philanthropist.

Innovation and networking as important factors in the performance of SMEs is affected by a country's business network approach. Currently, women have also dominated the business world evidenced by the customer's buying behavior established. With reference to the preceding sections on empirical research, it is quite clear that the subject of SMEs has been of interest to many researchers and consequently led to a number of researchers to be conducted SMEs. From the literature reviewed, there is unanimity in views held by different authors and researchers with regard to significance of SMEs in terms of the contribution to economies as well as challenges faced by these business organizations especially considering the role of women (Leach, 2015).

The literatures reviewed however indicate existence of knowledge gaps that warrant further research. For instance, previous studies indicate that women financial support is critical to the performance of the social development. Additionally, there is less empirical evidence on how capacity building, marketing support, and creation of business networks affect performance of small business owned by women in developing countries. Finally, most of these studies have given women-owned businesses located in rural setups a wide berth and concentrated on SMEs based in urban areas. For these reasons, the presented paper studies empowerment initiatives and performance of women-owned small and medium enterprises (SMEs) in rural setups in developing countries.

The relationship between the financial support, capacity building, marketing support and business networks and the success of women-owned businesses is a huge impact, which has a very significant role in social development. Financial support in the current article is characterized by affordable credit facilities and business grants. Capacity building is defined by training and skills development and infrastructural support. Marketing support and business networks and are characterized by sourcing for markets and, product promotion, and professional associations and linking with other women entrepreneurs in the rural areas, respectively. For centuries, theories have been used to explain, support, or rationalize practice in society. The increased economic independence for women in the society has had their share of these theories. In developed and developing countries, the support for women have expanded over the years and undergone reforms aimed at streamlining and rationalizing provision of equal opportunities (Leach, 2015).

9.4. EDUCATION FOR GIRLS

In developing countries, there is an increasing access to education for all children of school age. This is inclusive of the girl child who was secluded especially in the undeveloped or developing countries due to ill traditions. Traditional socio-cultural status of women have disadvantaged them and deprive women equal education opportunity, and source of wealth which directly and indirectly impacts negatively on their health status and life expectation, for example, in reaching health centers, in utilizing properly, in deciding where to seek medical services, in communicating with doctors etc. (Anderson et al., 2018).

In regards to all overall education, the current article recommends that education for girls be enhanced and be made a priority in policy formulation. This is in line with the study's finding that mortality decreased with an increase in educational attainment of the women especially that exhibited a negative effect on maternal mortality and morbidity. Girls in all countries should therefore be encouraged to attain at least a secondary level of education.

Previous studies have shown that younger and older maternal age has a deleterious effect on maternal mortality and morbidity. It also showed that the risk of mortality increased at higher parities, which indicates that women suffer and die because they are married as adolescents and are constrained into roles where their worth is defined by the number of children they bear. Too many children with short birth intervals and too late as revealed by this study is a pattern of childbearing that carries immense risks to the health of mothers which can be reduced by improving their socio-economic status. This can be achieved by policymakers if the educational costs for girls are reduced or by waiving school fees for girls who go beyond primary level to delay age at marriage especially in the rural areas. The implication of education for girls can result to women accessing better jobs to improve on their incomes and nutrition.

The educational planner finds himself in, in trying to satisfy both the social and manpower demand for education, under situations of limited resources. The different global governments have managed to implement primary education at the expense of deteriorating quality and increased public expenditure in the prevailing conditions of fast population growth rates. This is evident even in the Middle Eastern countries who for a long time secluded the role of women to being confined at the kitchen walls. It is important to understand to point out the dangers of rising population growth rates, on provision of necessary facilities, such as, health, education, food, and improvement of the standard of living in general (Anderson et al., 2018).

Different governments through the Ministry of Education have campaigned for the promotion of women education especially to higher levels. This is a crucial step in expanding provision of education in developing countries, which has substantial human resources. Consequently, this can be utilized effectively for socio-economic development through imaginative planning and appropriate educational reforms. Despite the academic controversies as to whether education contributes to economic development, in the minds of policy makers and technocrats education is perceived as an integral part of socio-economic development strategies.

Education reforms examining for women education has had a huge impact in the role that women get to play in terms of social development and well-being. These reforms have come in the wake of similar reforms in other developing countries in Sub-Saharan Africa, Latin America, and Asia, which have been tried with varying degrees of success. This has been in their attempts to cope with issues related to high rate of population growth, the growing level of unemployment and under-employment, the shortage of different types and levels of labor, equitable distribution of educational resources and the incidence of rural-urban migration. Reforms in Third World countries arose because of rapidly expanding educational systems which in turn escalated the cost of financing educational programs.

For centuries, theories have been used to explain, support, or rationalize practice in society. Educational systems were not a favor for most developing countries and this has been the primary bottleneck. Furthermore, the increased lack of sufficient support for the girl child has further declined the potential for development and each society has had their share of these theories. In developed and developing countries, educational systems have expanded over the years and undergone reforms aimed at streamlining and rationalizing educational provision (Blakemore et al., 2013).

Once society reaches consensus as to the needs and goals of education, school is used as an agency for transformation of society. Schools bring change through social and occupational differentiation. By helping to transform heterogeneous groups into well-informed citizens sharing social values, norms, and aspirations and developing national identities, education at the same time opens up avenues for social mobility. Through equality of opportunity, those able to benefit from the system excel as a result of their natural endowments. Differences in income, power, and status bestowed by education in some Way reflect individual abilities.

Since primary education is a basic human right and a prerequisite for socio-economic development, it is necessary for the nation to ensure that all those in the appropriate ages have access to the necessary educational facilities. This can be done through expansion of the facilities in conjunction with a suitable population program to ensure that the demand for the facilities does not outweigh the capacity of the society to supply. The education system should not be gender biased to avoid any prejudice since in this era; studies reveal the importance of equality and opportunity for all.

More school places should be made available yearly to cater for the increasing school-age population. Additionally, in the conservative rural areas, it is important to campaign and carry out initiatives addressing the importance of education for the girl child. For a society to have overall growth whether social or economic, it is crucial to eradicate poverty which can be achieved through education of all children equally. Short-term measures to ensure this would include extension of the already existing schools, particularly in the low-density zones (Northern and Southern Divisions) where there is relatively ample space for expansion (Blakemore et al., 2013).

10

MEN AND SOCIAL DEVELOPMENT

CONTENTS

10.1. INTRODUCTION

The role of men in the historical teachings is often biblical positioning men as the family providers. In regards to social development, men compared to women have been the main pioneers given the glass ceiling effect which has been very chaotic for the female gender. The major theoretical orientation in urban use of land was mainly based on the interplay of market forces also recognized that economic activities served its population especially for men who were the main sources of labor. Land was known to belong to men until in recent years in the promotion of gender equality. These activities he noted were profoundly influenced by the nature of that population and that population trends therefore influenced future land use in urban areas in various ways and degrees.

Men are also known as the centers providing services in functional areas such as: administration and protection; communication and protection; commerce and industry; social sciences and political power (Cornwall et al., 2012) (Figure 10.1).

Figure 10.1. Men have driven almost all aspects in social development taking up leadership positions and such progressive changes have represented a better life for the people e.g., the evolution of technological use.

Source: https://www.investors.com/news/economy/trump-bump-small-business-optimism-soars-to-12-year-high/.

Instances of stagnation and the short-lived failures occur when there is a technological change and a society has to get used to it. Short-lived poor economic productivity, inaccessibility to resources can also cause stagnation and failures. This development perspective can be effective by including environmental concerns in the management of the development process.

10.2. MEN'S MANAGERIAL ROLES AND DEVELOPMENT

Men have also taken up managerial roles. Management, as part of the development process, implies that development agenda in developing countries should include improved organizational capacity of local planning institutions, provisions in planning law and governance. These are the necessary pre-conditions of planning and management in the development process because the use of primary resources including land is central to them. They will lead to resolving the confusion surrounding the place of planning and management of urban land development in developing countries now that urbanization is the most challenging planning problem (Cornwall et al., 2012).

10.3. MEN'S POLITICAL PLATFORM AND DEVELOPMENT

On the political platform, men have seen the need to invest in social and physical infrastructure, and better management of energy supply also grows over the years. Throughout history, human societies have consistently influenced and been influenced by environmental processes resulting in conditions as we see them today. Consequently, the pristine environment has changed into the "cultural landscape" which denotes a transformed landscape due to the role of man in changing the face of the earth. Aspects of this include the emergence of secondary vegetation and ecological conditions where the original ones have been altered or destroyed; the rise of large-scale plantation system of agriculture in what was a jungle before; the creation of urban centers and their inherent infrastructure where rural communities predominated before, and so on.

The significance of the cultural landscape of developing countries and the role of the man has been a major influence on the economic activities of a region. There are also significant factors of development as they determine the sources of income and livelihood, type of employment and causes of unemployment and income-earning levels experienced within the target area# Resources determine required inputs for economic activities.

The economic activities are also influenced by population pressure and given that, it often grows steadily and is likely to have a lot of pressure on sources of employment, housing supply, and health and public health services in the town. This has implications for the town's future development. Prospects for solutions to this population expansion will be.-, mainly in efforts by employers to create more employment opportunities especially for the men, the development of diverse self-employment opportunities, and the informal sector. Most countries are privileged to be located favorably in an area with good accessibility and connectivity. This communication advantage has helped strengthen its economic/social activities and as a result, the provision of opportunities for the men has helped shape social development (Cornwall et al., 2012).

10.4. MEN'S ROLE IN GENERATING JOB OPPORTUNITIES

Men have taken the role of generating job opportunities for people. Urban growth is caused by decisions such as those to move the family or decisions to seek a new and better job or a more interesting social life. Educational attainment is a very important migration differential. This is especially so in rural-urban migration where the swelling numbers of job seekers (most of them men) compete for fist diminishing employment opportunities. Educational attainment is perhaps the most unstable migration differential because it changes with time as different time-scales demand varied levels of it (Cornwall et al., 2012) (Figure 10.2).

Figure 10.2. Men creating jobs.

Source: https://www.indiaspend.com/by-creating-green-jobs-indian-towns-can-save-the-environment-and-their-economies/.

The presence of relatives/friends includes the fact that they can be a source of psychological satisfaction; of economic. The support during the job-search process as well as initial settling on securing a job; and of reliable information about employment opportunities is very significant. Migration to the urban centers is a short-run solution to problems of population pressure as most of the migrants are on temporary jobs. There is the assumption that high population densities resulting in inaccessibility to agricultural land, land fragmentation and disputes, declining crop yields, food shortage, unemployment, and underemployment are 'push* factors for outmigration by the males to the prime centers in search for wage-earning jobs.

Similarly, the male adults may migrate to the prime urban centers in search of wage-earning jobs. The outmigration in search for employment outside the home does not have a strong influence as most of these migrants are casual laborers, earning small amounts of money almost no more than for their individual subsistence. The level of services is low and some centers lack some of the essential services. This is evident from the high rate of outmigration in search of alternative areas with jobs and adequate services. Men provide the regular unskilled workers with which the study has been concerned, but casual employees were predominantly females. It features prominently in discussions between trade unions and employing institutions regarding the need to improve housing and related conditions for the workers. In considerable population redistribution of farmers as well as migrant farmworkers, enhanced expansion of the industry and generated rural development hitherto unknown in many regions has led to increased social development (Cornwall et al., 2012).

REFERENCES

1. Andrea, C., Jerker, E., Alan, G., Chris, D., Chimaraoke, I., Akshay, K., et al., (2012). *Men and Development: Politicizing Masculinities* (p. 264). London: Zed Books Ltd.
2. Bashford, A., (2014). *Global Population: History, Geopolitics, and Life on Earth* (pp. 1–81). New York: Columbia University Press.
3. Benna, U. G., & Garba, S. B., (2016). *Population Growth and Rapid Urbanization in the Developing World* (p. 358). Pennsylvania: IGI Global.
4. Boll, A. M., (2007). *Multiple Nationality and International Law* (p. 626). Netherlands: Martinus Nijhoff Publishers.
5. Bonner, N., & David, W., (2013). *Antarctica* (pp. 21–73). New York: Elsevier.
6. Boserup, E., (2007). *Woman's Role in Economic Development* (p. 271). London: Earthscan.
7. Clare, H., William, G., Nissa, F., Alan, M., & Paul, N., (2013). *Population and Society* (pp. 8–71). London: SAGE.
8. Cummings, J. D., (2016). *Human Migration: Investigate the Global Journey of Humankind* (pp. 1–97). London: Nomad Press.
9. Deborah, K., Jonathan, G., & Bonnie, J. K., (2009). *Embracing Cultural Diversity: Australia: A Cross-Curricular Lesson Plan* (p. 13). New York: Lorenz Educational Press.
10. Donald, J. B., Douglas, L. A., & Richard, E. B., (2010). *The Population of the United States* (3rd edn., p. 704). New York: Simon and Schuster.
11. Doris, L. P., Sara, P., & Mokaya, B., (2016). *Diversity in African Languages: Selected Papers from the 46th Annual Conference on African Linguistics* (p. 600). Berlin: Language Science Press.

12. Drakakis-Smith, D., (2011). *Urbanization in the Developing World* (pp. 1–97). London: Routledge.
13. Dyson, T., (2013). *Population and Development: The Demographic Transition* (p. 288). London: Zed Books Ltd.
14. Eugenie, L. B., & Susan, M. W., (2011). *Global Urbanization* (p. 384). Pennsylvania: University of Pennsylvania Press.
15. Gould, W., (2015). *Population and Development* (pp. 18–65, 282). London: Routledge.
16. Hawthorn, G., (2013). *Population and Development: High and Low Fertility in Poorer Countries* (pp. 22–79). London: Routledge.
17. Judith, E. O. B., Sheri, A. B., & Lynn, S. L., (2013). *Gender Development* (p. 536). London: Psychology Press.
18. Leach, M., (2015). *Gender Equality and Sustainable Development* (pp. 1–85). London: Routledge.
19. Mangalam, J. J., (2015). *Human Migration: A Guide to Migration Literature in English 1955-1962* (p. 200). Kentucky: University Press of Kentucky.
20. Mokate, K. M., (2004). *Women's Participation in Social Development: Experiences from Asia, Latin America, and the Caribbean* (pp. 1–29). Jeddah: IDB.
21. Ness, I., (2014). *The Global Prehistory of Human Migration* (pp. 107–146). New Jersey: John Wiley & Sons.
22. OECD, (2010). *Trends in Urbanization and Urban Policies in OECD Countries What Lessons for China?: What Lessons for China?* (p. 219). Berlin: OECD Publishing.
23. Parrillo, V. N., (2015). *Diversity in America* (pp. 1–82). London: Routledge.
24. Prodi, R., (2002). *EU and Asia: Sharing Diversity in an Inter-Regional Partnership* (p. 26). Delhi: Institute of Southeast Asian Studies.
25. Richard, P. C., & Larry, J. G., (2011). *Human Population: Its Influences on Biological Diversity* (p. 244). Berlin: Springer Science & Business Media.
26. Rosset, E., (2017). *Aging Process of Population* (pp. 7–87). London: Elsevier.
27. Simon, J. L., (2014). *Population and Development in Poor Countries: Selected Essays* (p. 484). New Jersey: Princeton University Press.

28. Siwan, A., Lori, A. B., & Jean-Philippe, P., (2018). *Towards Gender Equity in Development* (p. 413). London: Oxford University Press.

29. Tamara, K. H., & Maris, A. V., (2015). *Family and Population in 19th Century America* (p. 266). New Jersey: Princeton University Press.

30. Thomas, E., Michail, F., Julie, G., Burak, G., Peter, J. M., Robert, I., McDonald, Susan, P., Maria, S., Marte, S., Karen, C. S., & Cathy, W., (2013). *Urbanization, Biodiversity and Ecosystem Services: Challenges and Opportunities: A Global Assessment* (p. 755). Berlin: Springer.

31. Wilson, R., (2018). *Meeting the Challenge of Cultural Diversity in Europe* (p. 240). London: Edward Elgar Publishing.

32. Wolfgang, L., William, P. B., Samir, K. C., & Samir, K. C., (2017). *World Population and Human Capital in the Twenty-First Century: An Overview* (p. 714). London: Oxford University Press.

INDEX

H

I

L

M

N

Printed in the United States
By Bookmasters